Land-Use Modelling in Planning Practice

The GeoJournal Library

Volume 101

For further volumes:
http://www.springer.com/series/6007

Eric Koomen · Judith Borsboom-van Beurden
Editors

Land-Use Modelling in Planning Practice

Springer

Editors
Eric Koomen
VU University Amsterdam
Spatial Economics/SPINlab
FEWEB/RE
De Boelelaan 1105
1081 HV Amsterdam
The Netherlands
e.koomen@vu.nl

Judith Borsboom-van Beurden
TNO Behavioural and Societal Sciences
PO Box 49, 2600 AA Delft
The Netherlands
judith.borsboom@tno.nl

ISSN 0924-5499
ISBN 978-94-007-1821-0 e-ISBN 978-94-007-1822-7
DOI 10.1007/978-94-007-1822-7
Springer Dordrecht Heidelberg London New York

Library of Congress Control Number: 2011935563

Printed on acid-free paper

Springer is part of Springer Science+Business Media (www.springer.com)

Foreword

A generation or more ago when land use transport models were first being developed, the focus was on how different models compared with one another in technical and theoretical terms. There was a concern for dynamics, for interaction and for calibration and validation, but less so for how such models might be interfaced with wider planning processes and the stakeholders that operate such systems and are indeed influenced by the plans that emerged from them. The state of the art then consisted of comparative studies of cross-sectional comprehensive spatial interaction model applications catalogued, for example, in the ISGLUTI Project – the International Study Group on Land Use Transport Interaction – and reported in the book by Webster, Bly and Paulley (1988). The dominant focus was very much in terms of the technical performance of models rather than their use in planning or policy-making.

As our experience of these models grew and evolved, this focus began to shift to the context in which models were best used. Onto the agenda came ideas about the various tools that had been developed to inform how we might best make good plans, and how these could be stitched together into coherent planning methods. Planning support systems in analogy to decision support in management were first formally suggested over 20 years ago by Britton Harris (1989) in his seminal article *Beyond geographic information systems: computers and the planning professional* as a way of bridging the development of computer models and tools with the activities of plan-making. Since then, a series of contributions to ways of building this bridge have been forged, the most recent being reported by Brail's (2008) in the collection of papers in his book *Planning Support Systems for Cities and Regions*. Many of these sketch the wider context and illustrate how a diversity of models and methods are coming together to define appropriate forums for dialogues between model builders, planners and the wider set of stakeholders involved in policy and its implementation.

So far we do not have a detailed blow by blow account of building and applying models as part of planning support systems. Until now that is, because this book represents the first such chronology of how a suite of land-use modelling tools called LUMOS – Land Use MOdelling System – which is centred on the Land Use Scanner model with another model Environment Explorer being sometimes used in parallel, is being fashioned to examine a wide array of different planning issues ranging from climate change to ways of reducing energy use in transport. This book should convince sceptics of the need to use formal tools in a sensitive and appropriate manner to explore different urban and regional futures that can best address the various grand challenges involving the environment that will dominate the next 20 years and beyond. All of the authors writing here provide a splendid picture of planning support systems in action, in fact of several variants of a generic planning support system fashioned around the various tools and models that have been developed by many groups in the Netherlands which are now maintained by PBL Netherlands Environmental Assessment Agency.

This book reflects experience of using these tools over a 15 year period from the time when the Land Use Scanner was first developed, through its development to finer scale levels of spatial resolution and through its development from an analytical tool to one with a direct optimisation capability. The first three chapters in the book set the context by describing these models and setting them in the wider context of spatial modelling more generally. Koomen, Hilferink and Borsboom-van Beurden provide a comprehensive and technically useful description in the first chapter where they define the basic structure and purpose of the model as a 'specification of regional demand for land, a definition of local suitability, an allocation module', and resulting depictions of future land use.

This introduction is followed by setting the LUMOS-models in an international perspective based on a report by Timmermans, Batty, Couclelis and Wegener who were involved in developing a critique of the experience in 2007. It might seem a little odd that one of these reviewers is writing this foreword but as a group, we had a privileged role in learning about the project, and thus I can communicate our feelings that this entire effort should be brought to the attention of the wider world of land-use and urban modellers as well as planners engaged in the search for good practice in the kinds of planning that LUMOS has been used to support. In fact, the LUMOS toolbox is unusual in that the models generally operate across several scales from the countrywide Netherlands itself down to quite small urban and rural regions. What indeed is impressive is the range of applications that are reported here. These pick up on significant questions about sprawl, environment, city compaction, climate change, and energy reduction in the context of sustainability. The toolbox does not quite extend to dealing with demographic factors per se but there are plenty of hooks to suggest how these other sectoral models can be plugged into any planning support system fashioned on LUMOS principles.

Before the various contributions move onto applications and extensions of the models, van Schrojenstein Lantman, Verburg, Bregt, and Geertman provide an interesting and informative review of land-use models ranging from land cover to cellular automata and thence to agent based models. They review six generic types in more detail including *GEOMOD2* which is land cover based, *SLEUTH* which is a cellular automata land development model, *UrbanSim* which is probably the best example of a contemporary land use transport model based on discrete choice theory but also embodying fine scale spatial grain with an agent-based focus, *IMAGE* an ecological–environmental framework to explore the long-term dynamics of global change, *CORMAS* a multi-agent framework simulating natural resources allocation, and *ILUMASS* a micro-simulation model of urban land use. A brief review of the modelling process involving calibration and validation is present and this sets the scene for many applications in practice.

Sustainability issues follow focusing on climate change – flooding risks and water damage, shortages and salt-water intrusion – biodiversity, accessibility and environmental impacts, quality of life, global business issues, and landscape quality. Transport modelling and its relation to land-use change are then explored using the example of a well-established model in which many features of the land market appear. This model which is called Tigris XL is linked to Land Use Scanner through the housing market which lies at the heart of linking different kinds of urban model. Applications then focus on the potential for resource allocation across the Netherlands and its region where the various simulation models are used to look at bio energy production and more general regional spatial strategy planning. Many important lessons for the use of models in planning support are gleaned from these analyses. These lessons have been noted many times but here, they are based on a wealth of experience which is only possible when you have had the sort of sustained modelling effort that has been characteristic of planning in the Netherlands for the last 20–30 years at least.

Future developments are then charted and it is here that we see how the critical mass built up from this experience provides an important guide to how these models might be extended and improved. Dekkers and Rietveld begin this process by developing a land market basis for Land Use Scanner while Kuijpers-Linde provides the wider context of planning support. Last but not least a new market-based land-use model is proposed by Borsboom-van Beurden and Zondag. This builds on the Tigris XL schema and from this and other contributions below, it clear that the general consensus is that all these tools need a stronger economic underpinning for the processes of land allocation that they simulate and forecast as well as optimise.

This book is a timely and important contribution to ways in which we might use models in planning, models in practice, and how we might best use them to inform the dialogue between professionals and decision-makers. Case studies are essential in this but all too often, we do not have enough detail to know how effective the models and tools applied have actually been. This book redresses this balance for it contains a wealth of experience that is not available anywhere else. What is unusual and impressive is the way this experience is being used to improve planning support, to reconcile a changing balance between experts, professionals, informed lay interests and the public-at-large. Joshua Epstein (2008) in a fascinating essay entitled *Why Model?* makes the point 'The choice, then, is not whether to build models; it's whether to build explicit ones. In explicit models, assumptions are laid out in detail, so we can study exactly what they entail'. The contributions in this book provide this explicitness that Epstein calls for in a way that provides us with clear rules of engagement for the use of models in planning.

London, UK Michael Batty

References

Brail, R. K. (Ed.). (2008). *Planning support systems for cities and regions*. Cambridge, MA: Lincoln Institute of Land Policy.

Epstein, J. M. (2008). Why model? *Journal of Artificial Societies and Social Simulation, 11*(4), 12. Available at http://jasss.soc.surrey.ac.uk/11/4/12.html. Retrieved March 2011.

Harris, B. (1989). Beyond geographic information systems: Computers and the planning professional. *Journal of the American Planning Association, 55*, 85–90

Webster, F. V., Bly, P. H., & Paulley, N. J. (1988). *Urban land use and transport interaction: Policies and models*. Farnborough, Hants: Gower.

Preface

Land-use modelling has been firmly established in Dutch planning practice in the past 10 years. The Land Use MOdelling System (LUMOS) toolbox, managed by PBL Netherlands Environmental Assessment Agency, has made this development possible. The development of this toolbox started in 1996 and builds on the cooperation of public research organisations such as PBL and the Agricultural Economics Research Institute (LEI), academic institutes (VU University Amsterdam) and commercial IT companies (Geodan and Object Vision). The ongoing model development process is unique in the sense that it reflects almost 15 years of continuous interaction between planners, researchers and IT specialists.

This book gives an overview of the wealth of recent applications and developments of the LUMOS toolbox. It contains contributions of the many partners that are active in applying and developing the toolbox and focuses specifically on the Land Use Scanner model that was applied in a wide range of policy-related studies in the past years. In addition to being employed for trend extrapolation, scenario studies and optimisation at the national level, the model has also been frequently used at the lower, regional scale level as is demonstrated in the various regional cases that are included in the book. Besides these applications, the book also considers some of the more theoretical aspects of land-use models and discusses various studies preparing the further development of the model. As such, this book is a continuation of the previous Dutch *Ruimtescanner* book published in 2001 that described the development and initial applications of Land Use Scanner.

The current book is aimed at planners and researchers worldwide that are interested in the current state of the art of land-use modelling in planning practice. It shows which types of applications are possible with current operational instruments and discusses possible pathways for further development. The book allows scholars and practitioners around the globe to learn from the extensive experience of Dutch planners and modellers. This may be particularly interesting since the Netherlands have a longstanding experience in this field, which is exemplified by the fact that the well-known and often-used CLUE and MOLAND-based models also originate here.

Obviously, the book is only a snapshot of work in progress. It does, for example, not document recent work related to climate adaptation that is carried out within the

Climate changes Spatial Planning and Knowledge for Climate research programs. Neither does it pay attention to the many recent land-use models – based on Land Use Scanner – that were built in international projects related to, amongst others, the catchment areas of the Rivers Elbe, Rhine and Meuse, the region of Flanders, Surinam and Honduras. It does also not highlight the pan-European EU-ClueScanner model commissioned by the European Commission that is built upon the Geo-DMS model framework underlying Land Use Scanner. This new model follows the specification of the Dyna-CLUE model and uses a dynamic version of the algorithm that is also underlying the new discrete version of Land Use Scanner. Publications on these and other new model developments can be found on the websites: www.lumos.info and www.feweb.vu.nl/gis/research/lucas.

The first part of the book discusses the scientific and theoretical aspects of applying land-use models. After a concise introduction of the Land Use Scanner model in Chapter 1, the evaluation of the two land-use models that comprise the LUMOS toolbox (Land Use Scanner and Environment Explorer) by an international audit committee in 2007 and their recommendations for improvement of the current models are summarised in Chapter 2. Following, Chapter 3 explores the theoretical foundation of current land-use models and examines the pros and cons of various concepts and methods in land-use modelling.

Then, Part II discusses a number of applications of Land Use Scanner for a wide range of research and policy questions in environment, agriculture and spatial planning, and at various scale levels. The Chapters 4 and 5 highlight the comprehensive application of Land Use Scanner at the national level for the Second Sustainability Outlook on the future of the Netherlands. Chapter 4 introduces this study, while Chapter 5 discusses the link to the Tigris XL transport model that was realised in order to be able to analyse the joint impact of spatial planning and transportation measures. Chapter 6 explores the potential for bio-mass production in a regional case-study in the Province of Friesland and its agro-economic benefits. Subsequently, the role of different optimisations of land-use patterns and their environmental impact in a regional spatial planning process in the Province of Overijssel is evaluated in Chapter 7. Then, Chapter 8 zooms in at the methodological aspects of a number of recent applications at the regional scale level and their similarities and dissimilarities.

The final part of the book reports recent research initiatives working towards the development of a new land-use model. Chapter 9 describes how information on actual land prices can be used to develop a new method for modelling land-use transitions in Land Use Scanner. Subsequently, in Chapter 10 the information needs of spatial planning, in particular on land-use changes, and the requirements to a new model from the perspective of actual policy questions, are considered. Lastly, the way forward to a model meeting those requirements, and the various options to realise such a model in a cost-efficient way, are outlined in Chapter 11.

This book would not have been here without the joint efforts of many individuals and organisations. We are particularly grateful to the authors who contributed to this book and the many people at PBL (notably Bas van Bemmel, Filip de Blois, Bart Rijken and Annemieke Righart) who helped with the logistics of production including the revision of text and graphics. In addition, we want to thank the Dutch National research programme 'Climate Changes Spatial Planning' for sponsoring part of the extensive work involved in editing the book.

We hope that this book provides inspiration to planners worldwide to use a modelling approach to better understand the spatial context of their planning problems and to suggest potential solutions. A demonstration version of the model has therefore been made available on a separate website (www.feweb.vu.nl/gis/landusescanner.htm) to familiarise users with the potential of this kind of tools.

Amsterdam, The Netherlands Eric Koomen
Delft, The Netherlands Judith Borsboom-van Beurden

Contents

Contributors

Michael Batty Centre for Advanced Spatial Analysis (CASA), University College London, 1-19 Torrington Place, London WC1E 6BT, UK, m.batty@ucl.ac.uk

Judith Borsboom-van Beurden TNO Behavioural and Societal Sciences, PO Box 49, 2600 AA Delft, The Netherlands, judith.borsboom@tno.nl

Arno Bouwman PBL Netherlands Environmental Assessment Agency, PO Box 303, 3720 AH Bilthoven, The Netherlands, arno.bouwman@pbl.nl

Arnold Bregt Laboratory of Geo-Information Science and Remote Sensing, Wageningen University, PO Box 47, 6700 AA Wageningen, The Netherlands, arnold.bregt@wur.nl

Arjen van der Burg Ministry of Infrastructure and Environment, PO Box 20901, 2500 EX The Hague, The Netherlands, arjen.vanderburg@minvrom.nl

Helen Couclelis Department of Geography, University of California, Santa Barbara, CA 93106, USA, cook@geog.ucsb.edu

Jasper Dekkers Department of Spatial Economics/SPINlab, VU University Amsterdam, De Boelelaan 1105, 1081 HV Amsterdam, The Netherlands, j.dekkers@vu.nl

Janneke van Dijk Agricultural Economics Research Institute (LEI), PO Box 29703, 7502 LS The Hague, The Netherlands, janneke.vandijk@ded.de

Egbert Dijk Province of Overijssel, PO Box 10078, 8000 GB Zwolle, The Netherlands, E.Dijk@overijssel.nl

Stan Geertman Faculty of Geosciences, Utrecht University, PO Box 80115, 3508 TC Utrecht, The Netherlands, s.geertman@geo.uu.nl

Karst Geurs Centre for Transport Studies, University of Twente, PO Box 217, 7500 AE Enschede, The Netherlands, k.t.geurs@utwente.nl

Maarten Hilferink Object Vision, c/o VU University Amsterdam, De Boelelaan 1087, 1081 HV Amsterdam, The Netherlands, mhilferink@objectvision.nl

Chris Jacobs Department of Spatial Economics/SPINlab, VU University Amsterdam, De Boelelaan 1105, 1081 HV Amsterdam, The Netherlands, c.g.w.jacobs@vu.nl

Arjen Koekoek Geodan, President Kennedylaan 1, 1079 MB Amsterdam, The Netherlands, arjen.koekoek@geodan.nl

Eric Koomen Department of Spatial Economics/SPINlab, VU University Amsterdam, De Boelelaan 1105, 1081 HV Amsterdam, The Netherlands; Geodan, President Kennedylaan 1, 1079 MB Amsterdam, The Netherlands, e.koomen@vu.nl

Tom Kuhlman Agricultural Economics Research Institute (LEI), PO Box 29703, 7502 LS The Hague, The Netherlands, tom.kuhlman@wur.nl

Marianne Kuijpers-Linde TNO Urban Development, PO Box 49, 2600 AA Delft, The Netherlands, marianne.kuijpers@tno.nl

Rienk Kuiper PBL Netherlands Environmental Assessment Agency, PO Box 30314, 2500 GH The Hague, The Netherlands, rienk.kuiper@pbl.nl

Jonas van Schrojenstein Lantman Nelen & Schuurmans, PO Box 1219, 3500 BE Utrecht, The Netherlands, jonas.vanschrojenstein@nelen-schuurmans.nl

Willem Loonen ProRail, PO Box 2038, F1.09, 3500 GA Utrecht, The Netherlands, willem.loonen@prorail.nl

Nga Phan-Drost Department of Spatial Economics/SPINlab, VU University Amsterdam, Amsterdam, The Netherlands, phandrost@gmail.com

Piet Rietveld Department of Spatial Economics, VU University Amsterdam, De Boelelaan 1105, 1081 HV Amsterdam, The Netherlands, p.rietveld@vu.nl

Harry Timmermans Urban Planning Group/EIRASS, Eindhoven University of Technology, PO Box 513, 5600 MB Eindhoven, The Netherlands, h.j.p.timmermans@bwk.tue.nl

Peter H. Verburg Institute for Environmental Studies, VU University Amsterdam, De Boelelaan 1087, 1081 HV Amsterdam, The Netherlands, peter.verburg@ivm.vu.nl

René Verburg Agricultural Economics Research Institute (LEI), PO Box 29703, 7502 LS The Hague, The Netherlands, rene.verburg@wur.nl

Michael Wegener Spiekermann & Wegener (S&W) Urban and Regional Research, Lindemannstrasse 10, D-44137 Dortmund, Germany, mw@spiekermann-wegener.de

Barry Zondag PBL Netherlands Environmental Assessment Agency, PO Box 30314, 2500 GH The Hague, The Netherlands, barry.zondag@pbl.nl

Part I
Introduction

Chapter 1
Introducing Land Use Scanner

Eric Koomen, Maarten Hilferink, and Judith Borsboom-van Beurden

1.1 Introduction

The PBL Netherlands Environmental Assessment Agency has a long tradition in land-use modelling. Indeed, the PBL has been putting spatially explicit models of land-use change into practice for almost 15 years The agency manages the Land Use MOdelling System (LUMOS) toolbox, which currently consists of two well-known models for simulating land-use change: Land Use Scanner and Environment Explorer; as well as a set of tools for pre- and post-processing of the modelling results, of the latter of which the Map Comparison Kit is an example.

Dealing with urban, natural and agricultural land functions all together, Land Use Scanner offers an integrated view of spatial changes in all types of land use. Since the development of its first version in 1997, it has been applied in a large number of policy-related research projects. These include the simulation of future land use following various scenarios (Borsboom-van Beurden, Bakema & Tijbosch, 2007; Dekkers and Koomen, 2007; Schotten and Heunks, 2001); the evaluation of alternatives for a new national airport (Scholten, Van de Velde, Rietveld & Hilferink, 1999); the preparation of the Fifth National Spatial Strategy (Schotten, Goetgeluk, Hilferink, Rietveld & Scholten, 2001); an outlook for the prospects of agricultural land use in the Netherlands (Koomen, Kuhlman, Groen & Bouwman, 2005); and the potential impact of climate change on land-use patterns (Koomen, Loonen & Hilferink, 2008). In addition to these Dutch applications, Land Use Scanner has also been used in several European countries (Hartje et al., 2008; Hoymann, 2010; Schotten et al., 2001; Wagtendonk, Julião & Schotten, 2001). For a full account of the methodological and technical details of the original model see Hilferink and Rietveld (1999). For an extensive overview of all publications related to Land Use Scanner, see www.lumos.info and www.feweb.vu.nl/gis.

E. Koomen (✉)
Department of Spatial Economics/SPINlab, VU University Amsterdam, De Boelelaan 1105, 1081 HV Amsterdam, The Netherlands; Geodan, President Kennedylaan 1, 1079 MB Amsterdam, The Netherlands
e-mail: e.koomen@vu.nl

E. Koomen, J. Borsboom-van Beurden (eds.), *Land-Use Modelling in Planning Practice*, The GeoJournal Library 101, DOI 10.1007/978-94-007-1822-7_1,

A brief overview of the various ways in which land-use models at the PBL have contributed to the information needed in the preparation of environmental and spatial planning policies is given in Section 1.2. The structure of the Land Use Scanner model is briefly discussed in Section 1.3. Finally, several implementation issues related to using Land Use Scanner in practice are highlighted in Section 1.4.

1.2 Land Use Scanner in the Context of Dutch Spatial Planning and Environmental Policy

The objective of most Dutch spatial planning-related Land Use Scanner applications is to provide probable spatial patterns of land-use change related to predefined conditions on demographic and economic scenario assumptions or specific policy interventions. Basically, three approaches can be followed:

1. elaboration of diverging scenarios;
2. extrapolation of trends;
3. optimisation of land use.

Depending upon the approach chosen, research or policy questions are translated into a regional demand for land, rules for allocation and a GIS-database, as is demonstrated in the descriptions of the following three applications.

1.2.1 Elaboration of Diverging Scenarios

To date, the majority of the applications of Land Use Scanner on a national level followed a scenario-based approach to deal with the uncertainties around future spatial developments. These uncertainties are largely determined by demographic and economic changes: population and GDP growth, ageing, decreasing household size, economic sector shifts and changes in productivity. Another major source of uncertainty is government intervention in the spatial domain. By describing a set of diverging views on the future – as is common in, for example, the reports of the Intergovernmental Panel on Climate Change (IPCC, 2001) – a broad range of spatial developments can be simulated, offering an overview of potential land-use changes. A scenario will, as such, not necessarily contain the most likely prospects, but, taken together, the simulations provide the bandwidth of possible land-use changes. In such a study, the individual scenarios should, in fact, not strive to be as probable as possible, but should stir the imagination and broaden the view on the future. Important elements are: plausible unexpectedness and informational vividness (Xiang & Clarke, 2003). An example of such a scenario-based simulation of land-use change is the Spatial Impressions project by the PBL (Borsboom-van Beurden et al., 2005; 2007). This analysis was performed to evaluate the possible impact on nature and landscape of economic and demographic changes in the future, as described in the first Sustainability Outlook study (MNP, 2004). The qualitative storylines of the original scenario framework were translated in spatially explicit assumptions, regarding the location preferences and future demand of a large number of land-use types, by means of expert-workshops and sector specific regional models. The results of the study were subsequently used to inform the National Parliament. The general public was also informed through, for example, publicity in the national media (Schreuder, 2005). The study pointed out that increased land use for housing, employment and leisure will contribute to significant further urbanisation, especially in the already heavily urbanised western part of the Netherlands. This will result in deterioration of nature areas and valuable landscapes, the extent depending upon the degree of government protection assumed in a scenario (Fig. 1.1).

Scenario-based, exploratory approaches as taken in the Spatial Impressions project follow a *what-if* approach: they indicate *what* may happen *if* certain conditions occur. This implies that the main task of the land-use model in these applications is not so much to create the most probable future land-use pattern, but rather to produce outcomes that investigate the range of possible land-use

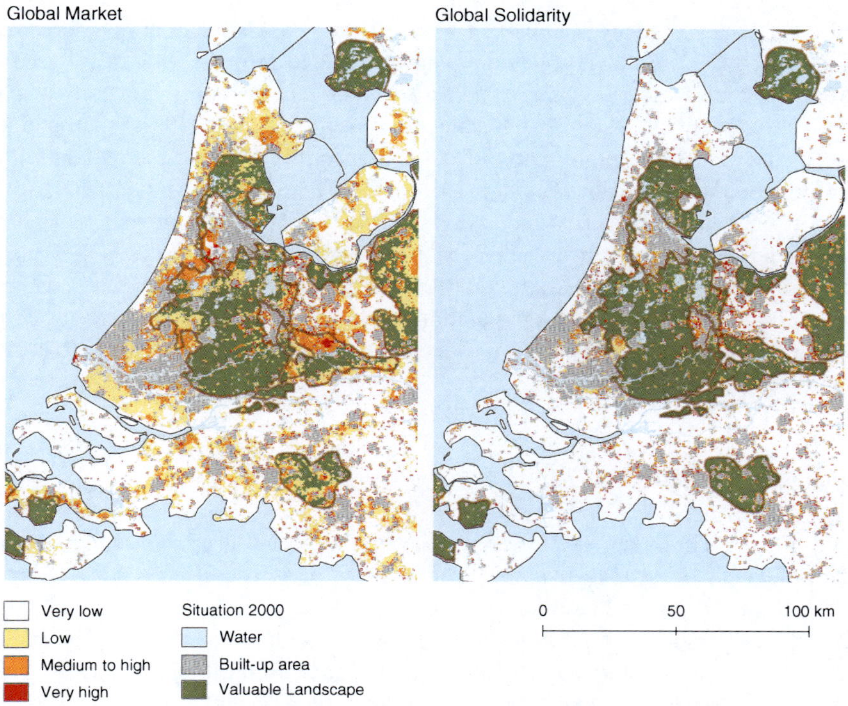

Fig. 1.1 Land use simulated according to the Global Market (*left*) and Global Solidarity (*right*) scenarios: the intensity of *red* areas indicates a possible increase in urban pressure; the *green* areas inside the *grey* contours signify valuable landscapes (Borsboom-van Beurden et al., 2005)

changes. The implementation of policy-specific scenarios that are used to sketch potential land-use patterns resulting from anticipated policies is a special type of scenario-based land-use simulation. This type of application is especially useful in impact assessment studies (see, for example, Chapter 7).

In practice, the provision of a range of possible scenario-based views of the future is often considered confusing by politicians and other decision-makers: they feel impelled to prefer a specific scenario, while the range of outcomes was meant, in the first place, to provide information on the magnitude of spatial changes and their interdependency with specific policies and interventions. Many decision-makers feel the need to have a 'business as usual' scenario, which can be considered as the most likely scenario. For that reason, more recent policy-oriented studies tend to put more emphasis on providing the most probable land-use patterns that reflect the extrapolation of current trends and policies.

1.2.2 Extrapolation of Trends

An example of extrapolation of trends is the Second Sustainability Outlook study for the Netherlands (MNP, 2007; PBL, 2010), which is described in detail in Chapter 4. This study analysed whether prevailing sustainability goals were being achieved and what policy objectives remained for the future. Social, economic and spatial trends that were current at that time were tracked to provide what is referred to as the Baseline Scenario. This version only takes into account policies that have been approved by the Dutch parliament or European Parliament. It does not include policy changes or the introduction of new policies, but assumes a continuation of prevailing policies. In this study, based on the Transatlantic Market scenario in the study Welfare, Prosperity and Quality of the Living Environment produced by the Dutch assessment agencies (CPB, MNP and RPB, 2006), average demographic and economic growth was assumed until 2040: more precisely a modest economic growth of 1.7% per year and a population growth to just over 17 million by 2040. This Baseline Scenario, representing average spatial pressure, is in line with the OECD baseline scenario (Fig. 1.2).

1.2.3 Optimisation of Land Use

Land-use modelling can also be applied to optimise land use from an ecological, environmental or spatial planning perspective, as is described in Chapter 7 and elsewhere in the literature (Loonen, Heuberger & Kuijpers-Linde, 2007). The Second Sustainability Outlook study also contains several examples of this approach: the study optimised land-use patterns according to six different policy themes or so-called viewpoints. For each of these themes, land use was simulated in such a way that specific, adverse developments were prevented.

The optimisation of land use for each policy theme started with an inventory of the autonomous developments that hamper the realisation of the current policy

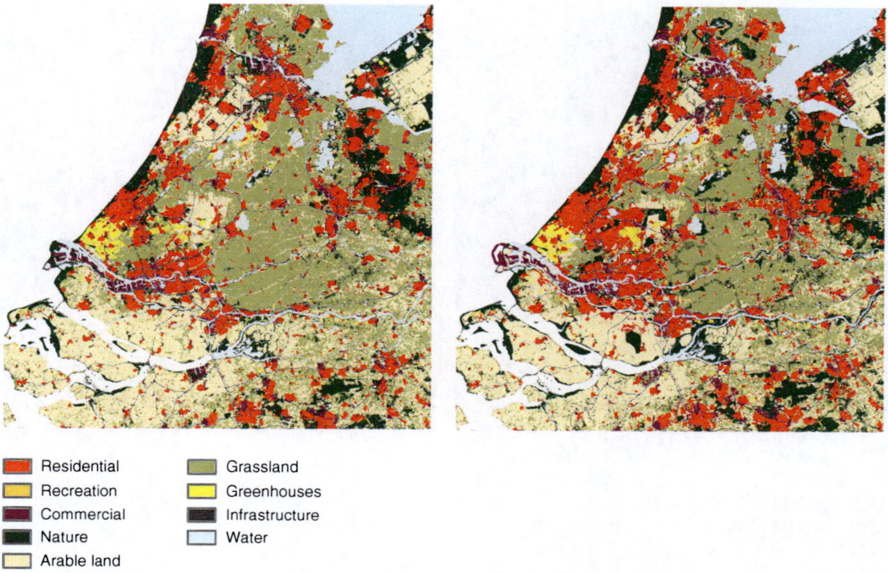

Fig. 1.2 Land use in base year 2000 (*left*) and in 2040 (*right*) according to the Baseline Scenario (Source: MNP, 2007)

objectives. This inventory is based on the trend-based simulation of land use, described in the preceding section. For the biodiversity theme, for example, the fragmentation of habitats through the construction of infrastructure (e.g. roads, railway lines) and the development of sites for housing and business parks are likely to lead to a lack of spatial coherence in the National Ecological Network and Natura2000 sites. What is more, the presence of agriculture and the lowering of the water-table cause pollution and drought, affecting the quality of nature areas. From the Robust Nature viewpoint, the projected spatial developments were, therefore, optimised according to specific planning objectives with the aim of displaying possible alternative land-use configurations that may result from policy interventions. The current Natura2000 sites formed the base for the optimisation. To avoid negative consequences on these Natura2000 sites and their species, buffer zones were designated to neutralise the environmental and hydrological impact of agricultural activities nearby. Further, areas with a high biodiversity were added to the Natura2000 sites on basis of either the occurrence of threatened and rare species or their adjacency to the Natura2000 sites.

After the optimisation step, it appeared that the total surface area for nature areas is about the same as for the Baseline Scenario, but it is much more geographically concentrated. As a result, the spatial preconditions for protected animal species improved considerably. An additional ecological assessment found that about 25% of the species had a better chance of sustainable preservation. Figure 1.3 shows the outcomes of the simulation of land use according to the principles of Robust Nature.

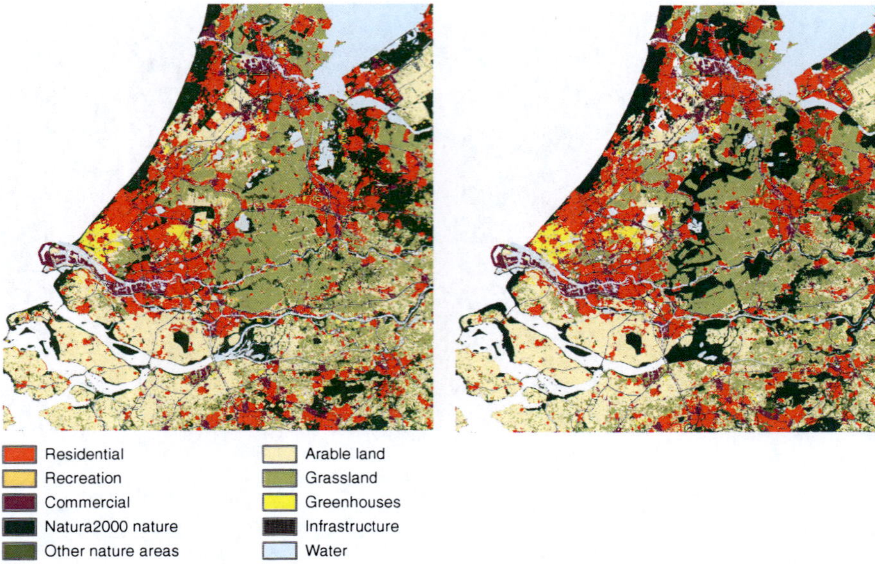

Fig. 1.3 Optimisation of land use in 2040 according the Baseline Scenario (*left*) and Robust Nature viewpoint (*right*)

1.2.4 Regional Applications

From the above, it follows that each approach has its own merits and supports the policy-making process in spatial planning in a different way by providing different information. But apart from these approaches, another distinction can be made relating to the scale of the study. With the exception of the study on the possible relocation of Amsterdam Airport (Scholten et al., 1999; Van de Velde et al., 1997) and a study for the Province of South-Holland (Borsboom-van Beurden et al., 2007; Bouwman, Kuiper & Tijbosch, 2006), all applications of Land Use Scanner up to 2007 have been performed at the national level. With the recent transfer of many responsibilities in spatial planning to the provinces, the need for information to support spatial planning at a regional level has increased. Besides, as is shown in Chapters 10 and 11, the current policy questions concerning spatial planning, environment and sustainability now require much more detailed information than was needed at the time Land Use Scanner was developed. This book contains a number of recent examples of the successful use of Land Use Scanner at a regional level (e.g. Chapters 7 and 8). For those who are interested in the technical aspects of the Land Use Scanner model, the general structure of the basic model and later versions is described in Section 1.3.

1.3 Model Structure

Land Use Scanner is a GIS-based model that simulates future land use through the integration of sector-specific inputs from other, dedicated models. The model is based on a demand-supply interaction for land, with sectors competing within suitability and policy constraints. It uses a comparatively static approach that simulates a future state in a limited number of time steps. Recent applications of the model simulate land-use patterns in three subsequent time-steps, each comprising one or more decades (MNP, 2007), whereas initial applications used only one or two time steps. Unlike many other land-use models, the objective of the Land Use Scanner is not to forecast the amount of land-use change, but rather to integrate and allocate future demand for land provided by different, external sources, such as specialised sector-specific models or policy intentions. This is shown in Fig. 1.4, which presents the basic structure of the Land Use Scanner model. The main components of this structure are discussed in the following subsections.

1.3.1 Regional Demand and Local Suitability

The basic structure of the model consists of a specification of regional demand for land, a definition of local suitability, an allocation module and resulting depictions of future land use. The first two of these components are described below. The two

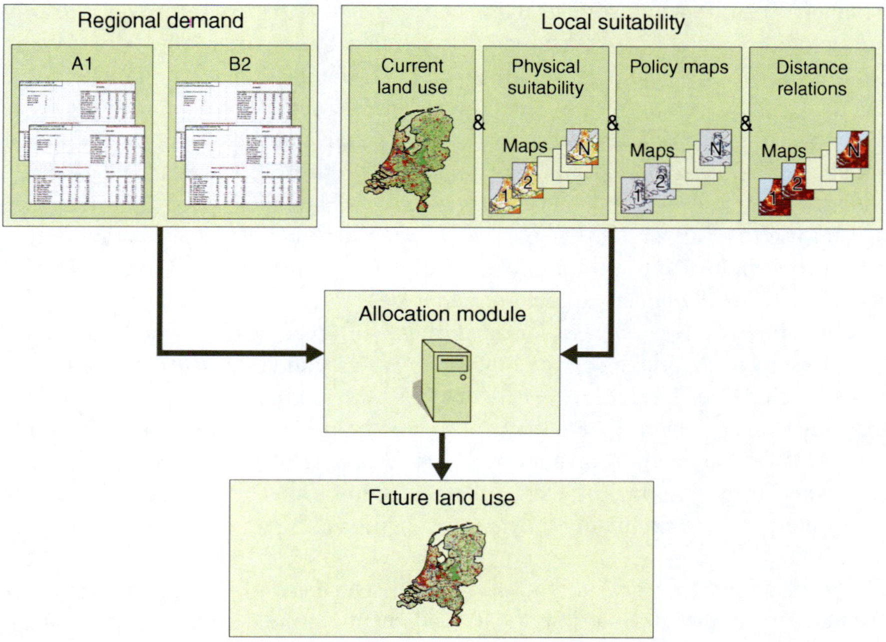

Fig. 1.4 Basic layout of the Land Use Scanner model

different allocation modules that are available in the model to simulate land-use patterns are introduced in the following subsections.

Regional Demand

External regional projections of the demand for land, which are usually referred to as land claims, are used as input for the model. These projections are specific for each land-use type and are derived from, for example, sector-specific models on housing or agriculture provided by specialised institutes or experts (when it comes to functions strongly dependant on policies, such as nature or water management). These projections of demand express for each land-use type the additional land demand. The total of the additional demand and the present area claimed by each land-use function is allocated to individual grid-cells based on the suitability of the cell for that particular land use.

Local Suitability

The definition of local suitability uses a large number of geo-datasets that refer to the following aspects: *current land use*, *physical properties*, *operative policies* and *market forces.*

Current land use is the starting point in the simulation of future land use. Various geo-datasets are used to construct a map of current land use in the base year of the simulation. Current land use is an important ingredient in the specification of both total regional demand for land and local suitability. For example, new housing is often located near to existing housing areas. However, because Land Use Scanner also allocates existing land use, current land-use patterns are not necessarily preserved in simulations. Transition costs can play an important role here, too, by preserving existing land use when that use is economically sound. The advantage of this flexibility is that dynamics in current land use can also be simulated, such as the conversion of obsolete business parks to new housing areas or the demolition of housing in regions with a shrinking population. This flexibility needs to be balanced with the geographical inertia that characterises especially the capital-intensive types of land use (e.g. urban land, greenhouses) and calls for sound information on the aspects that influence transition probability such as demolition costs. To date, this remains a relatively under-explored research area.

The *biophysical properties* of land (e.g. soil type and groundwater level) are especially important for the suitability specification of particular land-use types, such as in agriculture, where they directly influence possible yields, or for nature management, where they determine the possibilities of realising policy aims such as the creation of new wetlands. Biophysical properties are generally considered to be less important for urban functions, since the Netherlands has a long tradition of manipulating its natural conditions, in particular its hydrological conditions.

Operative policies, on the other hand, help steer Dutch land-use developments in many ways, and they are important components in the definition of suitability.

The designated zones of the National Ecological Network, where nature will be developed, or the municipal zoning plans are examples of spatial policies that stimulate the allocation of certain types of land use by enhancing its suitability. Conversely, policies can also reduce land suitability, through the definition of restrictions as is exemplified by various zoning laws related to, for example, groundwater protection and the preservation of landscape values.

The *market forces* that steer residential and commercial development, for instance, are generally expressed in distance relations to other, nearby land-use functions. Especially accessibility aspects such as proximity to railway stations, highway exits and airports are considered important factors that influence the location preferences of actors who are active in urban development. Other factors that reflect location preferences are, for example, the levels of service available from urban facilities or the attractiveness of the surrounding landscape.

The selection of the appropriate factors for all land-use types and their relative weighting are crucial steps in the preparation of the allocation of land uses and these largely determine the simulation outcomes. The relative weighting of the factors that describe the biophysical conditions, market forces and operative policies are normally assigned in such a way that they reflect the content of the particular trend, scenario or optimisation that is implemented land-use application.

1.3.2 Continuous Model

The original version of the Land Use Scanner model had a 500 m resolution with heterogeneous cells, each describing the relative proportion of all current land-use types. In this form it is referred to as a continuous model, since it uses a continuous description of the amount of land that is covered by each type of use in a cell. In the past, this approach has also been described as probabilistic, to reflect that the outcomes essentially describe the probability that a certain land use will be allocated to a specific location. This is different from most land-use models, which only describe one, dominant type of land use per cell.

The original, continuous model employs a logit-type approach, derived from discrete choice theory. Nobel prize winner McFadden made important contributions to this approach of modelling the choices made by actors between mutually exclusive alternatives (McFadden, 1978). In this theory, the probability that an individual selects a certain alternative is dependent on the utility of that specific alternative in relation to the total utility of all alternatives. This probability is, given its definition, expressed as a value between 0 and 1, although it will never reach either of these extremes. When translated into land use, this approach explains the probability of a certain type of land use at a certain location, based on the utility of that location for that specific type of use, in relation to the total utility of all possible uses.

The utility of a location can be interpreted as its suitability for a certain use. This suitability is a combination of positive and negative factors that approximate benefits and costs. The higher the utility or suitability for a land-use type, the higher the probability that the cell will be used for that type of use. Suitability is assessed by potential users and can also be interpreted as a bid price. After all, the user deriving the highest benefit from a location will offer the highest price. Furthermore, the model is constrained by two conditions, namely, the overall demand for each land-use function, and the amount of land that is available. By imposing these conditions, a doubly constrained logit model is established in which the expected amount of land in cell c that will be used for land-use type j is essentially described by the formula:

$$M_{cj} = a_j * b_c * e^{S_{cj}} \tag{1.1}$$

in which:

M_{cj} is the amount of land in cell c expected to be used for land-use type j;
a_j is the demand balancing factor (condition 1) that ensures that the total amount of allocated land for land-use type j equals the sector-specific claim;
b_c is the supply balancing factor (condition 2) that ensures that the total amount of allocated land in cell c does not exceed the amount of land that is available for that particular cell;
S_{cj} is the suitability of cell c for land-use type j based on its physical properties, operative policies and neighbourhood relations. The importance of the suitability value can be set by adjusting a scaling parameter.

The appropriate a_j values that meet the demand of all land-use types, are found in an iterative process, as is also discussed by (Dekkers & Koomen, 2007). This iterative approach simulates, in fact, a bidding process between competing land users (or, more precisely, land-use classes). Each use will try to get its total demand satisfied, but may be outbid by another category that derives higher benefits from the land. Thus, it can be said that the model, in a simplified way, mimics the land market. Governmental spatial planning policies that restrict the free functioning of the Dutch land market can be included in this process when they are interpreted as being either taxes or subsidies that cause an increase or decrease of the local suitability values respectively. In fact, the simulation process sort of produces shadow prices of land in the cells. This is discussed in more detail in the literature (Koomen & Buurman, 2002).

In reality, the process of allocating use is more complex than this basic description suggests. In brief, the most important extensions to the model are:

- The location of a selected number of land-use types (e.g. infrastructure, water) is considered as static and cannot be changed during simulations. Anticipated developments in these land-use types (e.g. the construction of a new railway line) are supplied exogenously to the simulations; that is they are directly included as simulation results and are not the derived from the iterative simulation process;

- The land-use claims are specified per region and this regional division may differ per land-use type, thus creating a more complex set of demand constraints;
- Minimum and maximum claims are introduced to make sure that the model is able to find a feasible solution. For land-use types with a minimum claim, it is possible to allocate more land. With a maximum claim it is possible to allocate less land. Maximum claims are essential if the total of all land-use claims exceeds the available amount of land;
- To reflect the fact that urban functions will, in general, outbid other functions at locations that are equally well suited for either type of land use, a monetary scaling of the suitability maps has recently been introduced (Borsboom-van Beurden et al., 2005; Groen, Koomen, Ritsema & Piek, 2004). In this approach, the maximum suitability value per land-use type is related to a realistic land price, ranging from, for example, 2.5 euros per square metre for nature areas to 35 euros per square metre for residential areas. The merits of this approach are currently being studied by others (Dekkers, 2005 and Chapter 9 this volume).

A more extensive mathematical description of the basic model and its extensions can be found in the literature (Hilferink & Rietveld, 1999).

The continuous model directly translates the probability that a cell will be used for a certain type of land use into an amount of land. A probability of 0.4 will thus, in the case of a 500 m × 500 m grid, translates into 10 ha. This straightforward approach is easy to implement and interpret but has the disadvantage of potentially providing very small surface areas for many different land-use types in a cell. This will occur especially if the suitability maps have little geographical variation in their values, a problem that can be solved by making the suitability maps more distinctive and pronounced. Another possible solution for this issue is the inclusion of a threshold value in the translation of probabilities into surface areas. Allocation can then be limited to those types of land use that, for example, have a probability of 0.2 or higher. The inclusion of such a threshold value calls for an adjustment of the allocation algorithm, to make sure that all land-use claims are met. This is feasible, however, and has been applied in the *Natuurplangenerator* model that aims to find an optimal spatial allocation of different types of nature within an area (Van Eupen & Nieuwenhuizen, 2002), which is in many ways similar to Land Use Scanner. Experience with this threshold value shows that insignificant quantities of land use are indeed set to zero, but if the threshold value is increased the model will have difficulties finding an optimal solution. This is due to the possibility that all probabilities are below the threshold value. Application of a threshold value in land-use simulation with Land Use Scanner remains to be tested and is a topic for further research.

For the visualisation of results, the simulation outcomes are normally aggregated and simplified in such a way that each cell portrays the single dominant category among a number of major categories. This simplification has, however, a substantial influence on the apparent results and may lead to a serious over-representation of some categories and an under-representation of others. To prevent the above mentioned issues, which are related to the translation and visualisation of the probability-related outcomes, an allocation algorithm was introduced that deals with homogenous cells – see the description of the discrete model in the following paragraphs.

1.3.3 Discrete Model

A revised version (4.7) of the Land Use Scanner model became available in 2005. This new version offered the possibility of using a grid of 100 m × 100 m, covering the terrestrial surface of the Netherlands in about 3.3 million cells. This resolution comes close to the size of actual building blocks and makes it possible to use homogenous cells that only describe the dominant land use. Furthermore, the revised version contained a new algorithm, which was developed to restrict calculation time. The algorithm finds the optimal allocation of land use for the given specified demand and suitability definition.

This new approach is referred to as the discrete model as it uses a discrete description of land use per cell: each cell is assigned only one type of land use from the total range of possible land-use types. Nowadays, the Land Use Scanner model has a flexible structure that allows for the selection of five different resolutions, ranging from 100 to 10,000 m, as well as the choice of using the discrete or continuous model, thus providing a total of 10 basic model variations.

The discrete allocation model allocates equal units of land (cells) to those types of land use that have the highest suitability, taking into account regional land-use demand. This discrete allocation problem is solved through a form linear programming, the solution of which is considered optimal when the sum of all suitability values corresponding to the allocated land use is maximal.

The allocation is subject to the following constraints:

- the amount of land allocated to a cell cannot be negative;
- in total, only 1 ha can be allocated to a cell;
- the total amount of land allocated to a specific land-use type in a region should be between the minimum and maximum claim for that region.

Mathematically the allocation problem can be formulated as:

$$\max_X \sum_{cj} S_{cj} X_{cj} \tag{1.2}$$

subject to:

$X_{cj} \geq 0$ for each c and j;

$\sum_j X_{cj} = 1$ for each c;

$L_{jr} \leq \sum_c X_{cj} \leq H_{jr}$ for each j and r for which claims are specified;

in which:

X_{cj} is the amount of land allocated to cell c to be used for land-use type j;
S_{cj} is the suitability of cell c for land-use type j;
L_{jr} is the minimum claim for land-use type j in region r; and
H_{jr} is the maximum claim for land-use type j in region r.

The regions for which the claims are specified may partially overlap, but for each land-use type *j*, a grid cell *c* can only be related to one pair of minimum and maximum claims. Since all of these constraints relate X_{cj} to one minimum claim, one maximum claim (which cannot be both binding) and one grid cell with a capacity of 1 ha, it follows that if all minimum and maximum claims are integers and that feasible solutions exist, the set of optimal solutions is not empty and lies between basic solutions in which each X_{cj} is either 0 or 1 ha.

The problem at hand is comparable to the well-known Hitchcock transportation problem that is common in transport-cost minimisation and, more specifically, the semi-assignment problem (Schrijver, 2003; Volgenant, 1996). The objective of the former problem is to find the optimal distribution in terms of minimised distribution costs of units of different homogenous goods from a set of origins to a set of destinations under constraints of a limited supply of goods, a fixed demand, and fixed transportation costs per unit for each origin – destination pair. The semi-assignment problem has the additional characteristic that all origin capacities are integers and that the demand for each destination is one unit of a specific homogenous good. Both are special cases of linear programming problems. The discrete allocation algorithm has two additional characteristics that are not incorporated in the mathematical formulation of the classical semi-assignment problem: (1) several (partially) overlapping regions are specified for the claims (although the regions of claims for the same land-use type may not overlap); and (2) it is possible to apply distinct minimum and maximum claims.

This mathematical problem, with its very large number of variables, calls for a specific, efficient algorithm. To improve the efficiency, a scaling procedure is applied and, furthermore, a threshold value is used. Scaling means that growing samples of cells are used in an iterative optimisation process that has proven to be fast (Tokuyama & Nakano, 1995). An optimisation is performed for each sample. After each optimisation, the sample is enlarged and the shadow prices in the optimisation process are updated in such way that the (downscaled) regional constraints continue to be met. To limit the number of alternatives under consideration, a threshold value is used: only allocation choices that are potentially optimal are placed in the priority queues for each competing claim. An important advantage of the algorithm used is that it enables an exact solution to be found with a desktop PC (Pentium IV-2.8 GHz, 1 GB internal memory) within several minutes, provided that feasible solutions exist and all suitability maps have been prepared in an initial run of the model. Running the model for the first time takes just over an hour as all base data layers have to be constructed. These data sets are then stored in the application files (in a temporary folder) to speed up further calculation.

The constraints that are applied in the new discrete allocation model are equal to the demand and supply balancing factors applied in the original, continuous version of the model. In fact, all the extensions to the original continuous model related to the fixed location of certain land-use types, the use of regional claims, the incorporation of minimum/maximum claims, and the monetary scaling of the suitability maps also apply to the discrete model. Similar to the original model, the applied optimisation algorithm of the discrete model aims to find shadow prices for the regional demand constraints that increase or decrease the suitability values, such that the allocation based on the adjusted suitability values corresponds to the regional claims. The main difference of the discrete model from the continuous model is that each cell only has one land-use type allocated; meaning that for each land-use type the share of allocated land is zero or one. From a theoretical perspective the models are, however, equivalent if the scaling parameter that defines the importance of the suitability values becomes infinitely large. In that case the continuous model would also strictly follow the suitability definition in the allocation and would produce homogenous cells. This procedure is, however, theoretical and cannot be applied in the calculations due to computational limitations. A more extensive discussion of the two available algorithms and an assessment of their performance is described in a separate report on calibration (Loonen & Koomen, 2009).

1.4 Land Use Scanner in Practice: Implementation Issues

With the model structure clarified, it is time to discuss the practical aspects of land-use modelling. For the PBL it is not so much the outcomes of the land-use simulations themselves that are of interest, but much more the meaning of these outcomes for assessing the environmental, ecological and spatial impact of land use. In fact, Land Use Scanner is part of a larger model chain. Its input is derived from sector-specific models that provide the future demand for land, whereas its output is used in specialised ecological, environmental or hydrological models to assess specific impacts. Land Use Scanner thus bridges these different model components (see Fig. 1.5), meaning that pre- and post-processing of data play an important role in the entire model chain. This section discusses the five main activities that have to be carried out to implement a new application in Land Use Scanner.

Construction of a Base Map of Land Use and Classification of Land Use

Firstly, a base map has to be created which contains data for all distinguished land-use types. For the Netherlands, this map is usually based on the latest versions of the datasets Land Use Statistics from Statistics Netherlands and the National Land Use Database (CBS, 2002; Thunissen & De Wit, 2000). The classification of land-use types should be in line with the definitions used by the sector models that simulate the demand for land. Often this raises questions on the translation of activities and objects, such as employment, farms and houses, to land use: are small

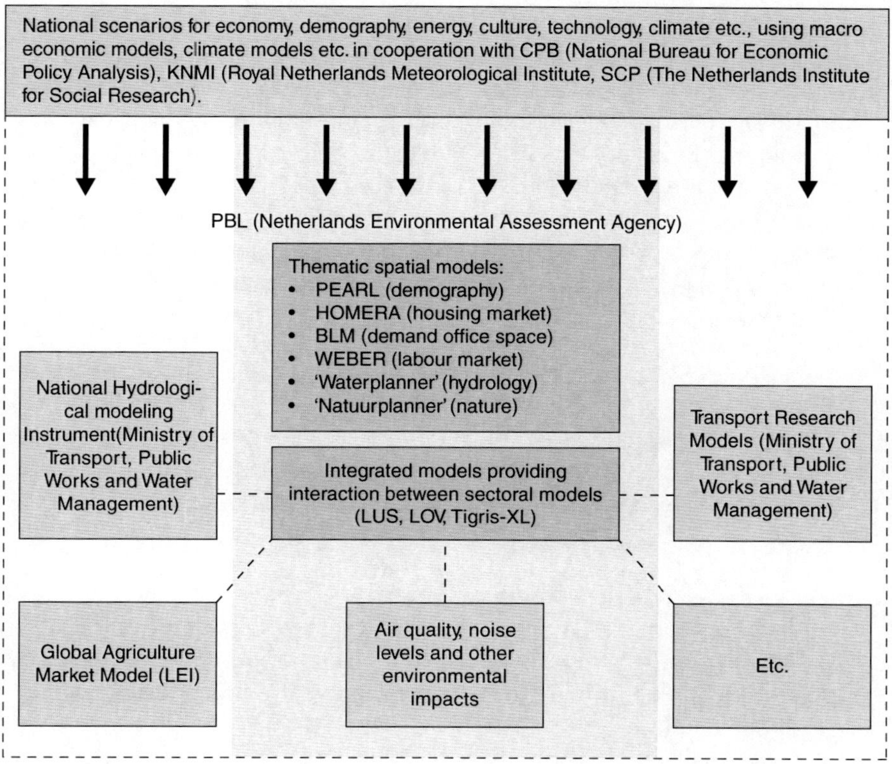

Fig. 1.5 The impact assessment modelling chain

roads, ditches, playgrounds also to be included? How to translate square meters of floor space to business parks? For most simulations, between 20 and 35 different categories of land use are distinguished.

Collecting the Demand for Land

For all land-use types, data on their future demand for land have to be gathered. For some types of land use this can be done relatively easily, by deriving data from specialised models, such as those for the housing market and employment or from agro-economic models. Often, however, definitions, spatial units, explanatory theories and underlying assumptions do not match and temporary solutions have to be found (Dekkers & Koomen, 2006). This requires quite some pre-processing. Furthermore, it is more difficult to find reliable data on the demand for land for sectors that are not so market-driven, such as nature, outdoor leisure and water management. For these sectors, one often has to fall back on the expected effects on future land use of policies, such as the realisation of the National Ecological Network or groundwater protections coming from experts.

Collection of Geo-Datasets Representing Specific Aspects of Suitability

Once all data on the demand for land have been acquired, maps related to specific aspects of suitability of grid cells for a particular type of land use need to be collected. In the course of time, the model configuration has become more and more elaborate and refined as each new study with the model was used to build upon and improve its previous configuration. The most recent model configuration contains more than 400 geo-datasets, which are regularly updated. An important element is the planning of new residential, business and nature development locations, because the probability that a certain land-use type will be realised here is very high.

Set-Up of Allocation Rules and Attachment of Relative Weights

For all land-use types, allocation rules describing the relevant suitability aspects maps and their relative weights are assigned using a scripting language (Data Model Server). Weights are attached that correspond with the relative importance of a particular aspect for the overall suitability definition for a particular land-use type. The more detailed the classification, the more time-consuming this part of the modelling is.

Conversion and Assessment of Model Outcomes

After land use has been simulated according to these steps, the outcomes can be presented as land-use maps. The results can also be used for the calculation of land-use based indicators within the Land Use Scanner itself (Bubeck & Koomen, 2008; Ritsema van Eck & Koomen, 2008). These indicators highlight specific aspects of the results, such as locations of changed land use, impacts on natural areas and valuable landscapes and various urbanisation processes. To facilitate the further exploration of these results and help with their interpretation tools such as the Map Comparison Kit (Visser & De Nijs, 2006) exist. In addition, more complex indicators related to environmental and spatial quality can be obtained by using the outcomes in specialised ecological, environmental or hydrological models. Often considerable post-processing is needed to link land-use simulation outcomes to these specialised models because, here too, definitions, spatial units, theories and assumptions are not harmonised. In 2009, the calculation of flooding risks was significantly improved by coupling the DamageScanner model (Van der Hoeven, Aerts, Van der Klis & Koomen, 2008) to Land Use Scanner. Even so, this end of the model chain needs considerable amelioration. An additional option for visualising and interpreting results is the construction of elaborate three-dimensional representations of changed land use, an approach that has been pursued in different research projects (Borsboom-van Beurden, Van Lammeren & Bouwman, 2006; Lloret, Omtzigt, Koomen & De Blois, 2008).

From this brief overview of implementation issues, it follows that the quality of the outcomes is mainly determined by the quality of the entire model chain, in which the actual allocation of land use is only one of the many issues. When using Land Use Scanner, special attention needs to be paid to the coherence and consistency existing between sector specific models and the Land Use Scanner model.

References

Borsboom-van Beurden, J. A. M., Bakema, A., & Tijbosch, H. (2007). A land-use modelling system for environmental impact assessment; Recent applications of the LUMOS toolbox. Chapter 16. In E. Koomen, J. Stillwell, A. Bakema, & H. J. Scholten (Eds.), *Modelling land-use change; Progress and applications* (pp. 281–296). Dordrecht: Springer.

Borsboom-van Beurden, J. A. M., Boersma, W. T., Bouwman, A. A., Crommentuijn, L. E. M., Dekkers, J. E. C., & Koomen, E. (2005). *Ruimtelijke Beelden; Visualisatie van een veranderd Nederland in 2030*. RIVM report 550016003. Bilthoven: Milieu- en Natuurplanbureau.

Borsboom-van Beurden, J. A. M., Van Lammeren, R. J. A., & Bouwman, A. A. (2006). Linking land use modelling and 3D visualisation: A mission impossible? In J. Van Leeuwen & H. Timmermans (Eds.), *Innovations in design and decision support systems in architecture and urban planning* (pp. 85–102). Dordrecht: Springer.

Bouwman, A. A., Kuiper, R., & Tijbosch, H. (2006). *Ruimtelijke beelden voor Zuid-Holland*. Rapportnummer 500074002.2006. Bilthoven: Milieu- en Natuurplanbureau.

Bubeck, P., & Koomen, E. (2008). *The use of quantitative evaluation measures in land-use change projections; An inventory of indicators available in the land use scanner. Spinlab Research Memorandum SL-07*. Amsterdam: Vrije Universiteit Amsterdam/SPINlab.

CBS (2002). *Productbeschrijving Bestand Bodemgebruik*. Voorburg: Centraal Bureau voor de Statistiek.

CPB, MNP and RPB (2006). *Welvaart en Leefomgeving. Een scenariostudie voor Nederland in 2040*. Den Haag: Centraal Planbureau, Milieu- en Natuurplanbureau en Ruimtelijk Planbureau.

Dekkers, J. E. C. (2005). *Grondprijzen, geschiktheidkaarten en parameterinstelling in de Ruimtescanner*. Technisch achtergrondrapport bij Ruimtelijke Beelden. MNP report 550016005. Bilthoven: Milieu- en Natuurplanbureau.

Dekkers, J. E. C., & Koomen, E. (2006). *De rol van sectorale inputmodellen in ruimtegebruiksimulatie; Onderzoek naar de modellenketen voor de LUMOS toolbox. SPINlab research memorandum SL-05*. Amsterdam: Vrije Universiteit Amsterdam.

Dekkers, J. E. C., & Koomen, E. (2007). Land-use simulation for water management: Application of the land use scanner model in two large-scale scenario-studies. Chapter 20. In E. Koomen, J. Stillwell, A. Bakema, & H. J. Scholten (Eds.), *Modelling land-use change; progress and applications* (pp. 355–373). Dordrecht: Springer.

Groen, J., Koomen, E., Ritsema, van E. J., & Piek, M. (2004). *Scenario's in kaart; model- en ontwerpbenaderingen voor toekomstig ruimtegebruik*. NAi Uitgevers/Ruimtelijk Planbureau, Rotterdam/Den Haag.

Hartje, V., Ansmann, T., Blazejczak, J., Gomann, H., Gornig, M., Grossman, M., et al. (2008). Szenarioanalyse der Regionalisierung der Driving Forces und Pressures des globalen Wandels in einem mittleren Flusseinzugsgebiet. Chapter 2. In: Wirkungen des globalen Wandels auf den Wasserkreislauf im Elbegebiet – Risiken und Optionen. Schlussbericht zum BMBF-Vorhaben GLOWA-Elbe II.

Hilferink, M., & Rietveld, P. (1999). Land use scanner: An integrated GIS based model for long term projections of land use in urban and rural areas. *Journal of Geographic Systems, 1*(2), 155–177.

Hoymann, J. (2010). Spatial allocation of future residential land use in the Elbe river basin. *Environment and Planning B: Planning and Design, 37*(5), 911–928.

IPCC (2001). *Climate change 2001: Synthesis report. A contribution of Working Groups I, II and III to the Third Assessment report of the Intergovernmental Panel on climate change*. Cambridge: Cambridge University Press.

Koomen, E., & Buurman, J. J. G. (2002). Economic theory and land prices in land use modeling. In M. Ruiz, M. Gould, & J. Ramon (Eds.), *5th AGILE conference on geographic information science proceedings* (pp. 265–270). Palma (Illes Balears), Spain: Universitat de les Illes Balears.

Koomen, E., Kuhlman, T., Groen, J., & Bouwman, A. A. (2005). Simulating the future of agricultural land use in The Netherlands. *Tijdschrift voor Economische en Sociale Geografie (Journal of Economic and Social Geography), 96*(2), 218–224.

Koomen, E., Loonen, W., & Hilferink, M. (2008). Climate-change adaptations in land-use planning; A scenario-based approach. In L. Bernard, A. Friis-Christensen, & H. Pundt (Eds.), *The European information society; Taking geoinformation science one step further* (pp. 261–282). Berlin: Springer.

Lloret, J. R., Omtzigt, N., Koomen, E., & De Blois, F. (2008). 3D visualisations in simulations of future land use: Exploring the possibilities of new, standard visualisation tools. *International Journal of Digital Earth, 1*(1), 148–154.

Loonen, W., Heuberger, P., & Kuijpers-Linde, M. (2007). Spatial optimisation in land-use allocation problems. Chapter 9. In E. Koomen, J. Stillwell, A. Bakema, & H. J. Scholten (Eds.), *Modelling land-use change; progress and applications* (pp. 147–165). Dordrecht: Springer.

Loonen, W., & Koomen, E. (2009). *Calibration and validation of the land use scanner allocation algorithms*. Bilthoven: PBL-report. Netherlands Environmental Assessment Agency.

McFadden, D. (1978). Modelling the choice of residential location. In A. Karlqvist, L. Lundqvist, F. Snickars, & J. W. Weibull, (Eds.), *Spatial interaction theory and planning models* (pp. 75–96). Amsterdam: North Holland Publishers.

MNP (2004). *Kwaliteit en toekomst. Verkenning van de duurzaamheid*. RIVM report. Bilthoven: Milieu- en Natuurplanbureau.

MNP (2007). *Nederland Later; Tweede Duurzaamheidsverkenning deel fysieke leefomgeving Nederland*. MNP-publicatienr.500127001/2007. Bilthoven: Milieu- en Natuurplanbureau.

PBL (2010). *The Netherlands in the Future. Second Sustainability Outlook: The physical living environment in The Netherlands*. Bilthoven: Netherlands Environmental Assessment Agency (PBL).

Ritsema, van E. J., & Koomen, E. (2008). Characterising urban concentration and land-use diversity in simulations of future land use. *Annals of Regional Science, 42*(1), 123–140.

Scholten, H. J., Van de Velde, R. J., Rietveld, P., & Hilferink, M. (1999). Spatial information infrastructure for scenario planning: The development of a land use planner for Holland. In J. Stillwell, S. Geertman, & S. Openshaw (Eds.), *Geographical information and planning* (pp. 112–134). Berlin/Heidelberg/New York: Springer.

Schotten, C. G. J., Goetgeluk, R., Hilferink, M., Rietveld, P., & Scholten, H. J. (2001). Residential construction, land use and the environment. Simulations for The Netherlands using a GIS-based land use model. *Environmental Modeling and Assessment, 6*, 133–143.

Schotten, C. G. J., & Heunks, C. (2001). A national planning application of Euroscanner in The Netherlands. Chapter 17. In J. C. H. Stillwell & H. J. Scholten (Eds.), *Land use simulation for Europe* (pp. 245–256). Amsterdam: Kluwer Academic Publishers.

Schotten, C. G. J., Heunks, C., Wagtendonk, A. J., Buurman, J. J. G., de Zeeuw, C. J., Kramer, H., Boersma, W. T. (2001). Simulating Europe in the 21th century. NRSP-2 report 00-22. BCRS, Delft.

Schreuder, A. (2005). Doemscenario's over natuurbehoud. *NRC-Handelsblad January, 15*, p. 2.

Schrijver, A. (2003). *Combinatorial optimization – polyhedra and efficiency*. Berlin: Springer.

Thunissen, H. A. M., & De Wit, A. J. W. (2000). *The National land cover database of the Netherlands*. Amsterdam: ISPRS XXXIII.

Tokuyama, T., & Nakano, J. (1995). Efficient algorithms for the Hitchcock transportation problem. *SIAM Journal on Computing, 24*(3), 563–578.

Van de Velde, R. J., Schotten, C. G. J., Van der Waals, J. F. M., Boersma, W. T., Ouwersloot, H., & Ransijn, M. (1997). Ruimteclaims en ruimtelijke ontwikkelingen in de zoekgebieden voor de toekomstige nationale luchtinfrastructuur (TNLI). Quickscan met de Ruimtescanner. RIVM-rapport 711901024. RIVM, Bilthoven.

Van der Hoeven, E., Aerts, J., Van der Klis, H., & Koomen, E. (2008). An integrated discussion support system for new Dutch flood risk management strategies. Chapter 8. In S. Geertman &

J. C. H. Stillwell (Eds.), *Planning support systems: Best practices and new methods* (159–174). Berlin: Springer.

Van Eupen, M., & Nieuwenhuizen, W. (2002). *NatuurPlanGenerator: Gebruikershandleiding (Versie 1.0). Werkdocument Natuurplanbureau*. Wageningen: Alterra.

Visser, H., & De Nijs, T. (2006). The map comparison kit. *Environmental Modelling & Software, 21*(3), 346–358.

Volgenant, A. (1996). Linear and semi-assignment problems: A core oriented approach. *Computers and Operations Research, 23*(10), 917–932.

Wagtendonk, A. J., Julião, R. P., & Schotten, C. G. J. (2001). A regional planning application of Euroscanner in Portugal. Chapter 18. In J. C. H. Stillwell & H. J. Scholten (Eds.), *Land use simulation for Europe* (pp. 257–291). Amsterdam: Kluwer.

Xiang, W. N., & Clarke, K. C. (2003). The use of scenarios in land-use planning. *Environment and Planning B, 30*, 885–909.

Chapter 2
Lumos Models from an International Perspective

Harry Timmermans, Michael Batty, Helen Couclelis, and Michael Wegener

2.1 Introduction

This chapter summarises the main findings of an international scientific audit of the Land Use Scanner and Environment Explorer models as described in the report of the audit committee (Timmermans, Batty, Couclelis & Wegener, 2007). Both models are part of the LUMOS toolbox maintained by the PBL Netherlands Environmental Assessment Agency. The audit was based on a selection of documents (reports, conference papers and journal articles) related to these models, and a site visit at PBL that took place from 8 to 9 January 2007, during which a series of presentations about and software demonstrations of the models were given.

The aim of the audit committee was to evaluate the scientific quality of the processes and products of PBL relating to the application of land-use models, primarily at the national level. The two models are used to help provide integrated environmental policy assessments and exploratory studies for the Netherlands. Because PBL is increasingly involved in regional and local land-use assessments, the appropriateness of the models for use at lower spatial scales and finer resolutions was also considered. In addition, the audit committee was asked to advise on improving scientific quality in future model development and use, and on actions to be taken to ensure an internationally prominent role for PBL in the use of land-use modelling to support environmental policy-making. The auditors also addressed PBL's mission and the resources available to it for the fulfilment of that mission.

More specifically, the audit committee posed the following questions:

1. Does PBL use the most appropriate model type(s) and suite(s) to produce the information needed for assessments and exploratory studies?
2. Are the resources available to PBL adequate and appropriate for the sort of use it makes of land-use models?

H. Timmermans (✉)
Urban Planning Group/EIRASS, Eindhoven University of Technology, PO Box 513, 5600 MB Eindhoven, The Netherlands
e-mail: h.j.p.timmermans@bwk.tue.nl

E. Koomen, J. Borsboom-van Beurden (eds.), *Land-Use Modelling in Planning Practice*, The GeoJournal Library 101, DOI 10.1007/978-94-007-1822-7_2,

3. Are the results of land-use model applications for environmental impact assessment adequately used in policy report making processes inside PBL, such as Outlooks and Balances, and in joint projects with other, relevant national and international agencies?
4. Are the results of studies using land-use models reported in a way that communication with policy-makers is well targeted and optimised, while maintaining an acceptable level of scientific quality?
5. Can the auditors advise PBL on its data, model and application methods for the coming years?

2.2 Audit Committee's Findings and Recommendations

2.2.1 Appropriateness of the Models

Does PBL use the most appropriate model type(s) and related software for pre- and postprocessing to produce the information needed for assessments and exploratory studies?

Both Land Use Scanner and Environment Explorer simulate the mechanisms that balance changes in the demand for and supply of various types of land use. Each model classifies these land uses in slightly different ways, but both are based on the same underlying data set. Land Use Scanner is based on principles taken from utility and discrete choice theory, whereas Environment Explorer is based on constrained cellular automata, reflecting simple decision rules associated with land development.

Land Use Scanner, originally developed in 1997, has made a significant contribution to the state of the art of cell-based land-use modelling. The discrete choice (logit) model for land-use allocation on which Land Use Scanner is based allows users to test different assumptions about the degree of rationality of choice in behaviour of actors driving land-use changes, ranging from full rationality through bounded rationality to random choice. In addition, a multitude of extensions of the basic model, including models of strategic behaviour, are possible. In the first, continuous version of the model (1997), land uses could be represented simultaneously as a proportion of the total land per land unit. In the second, discrete version of the model (2005), only one land-use type is represented per land unit. Both versions of the model can process grid cells of 500 m × 500 m and 100 m × 100 m. Allocation of land uses to cells in the continuous model occurs using bi-proportional adjustment such that demand for land by each land-use type and supply of land in each cell are balanced (or remain above or below these constraints). Allocation of land uses to cells in the discrete model occurs by optimisation (by finding an optimal solution of the Linear Semi Assignment Problem). The calibration of the continuous version of the model was originally done by iterative search for best-fit coefficients of the discrete choice model. Recent calibration methods for both versions use logistic regression. There was some concern within the auditing committee as to whether this calibration method is consistent because it does not take account of the versions' regional constraints.

Environment Explorer, also originally developed in 1997 and based on earlier work by White and Engelen, is one of the more advanced cellular automata models of land-use change used in practice. Environment Explorer is a two-level model: at a macro level it allocates demand for land by sector to 40 economic regions (COROP regions); and at a micro level it allocates regional demand to 500 m $\times$ 500 m grid cells in which only one land-use type is represented. The micro level of Environment Explorer is more comprehensive than that of Land Use Scanner as it is linked to a four-step transport model (trip generation, trip distribution, modal split and assignment) and an environmental impact model of traffic-generated noise and air pollution. It is not clear whether these models are state of the art, i.e. whether the transport model is able to model congestion or elastic demand. Land use is allocated to each grid cell via transition/decision rules based on intrinsic suitability, zoning, accessibility, neighbourhood effects and a random factor. Calibration of the micro-model is achieved by a pattern-matching algorithm that iteratively adjusts the many coefficients of the micro-model to maximise goodness of fit. The core of the model is at the cellular automata level, with the model being extended to embrace models at greater spatial scales within a more general process of integrated assessment.

The fact that Land Use Scanner and Environment Explorer are being used by PBL in much the same way implies that in actual applications the differences between them have become smaller over the years. Both can now use the same data and practically the same classification of land uses. They both deal with land-use allocation, and in certain cases the main difference between them is just the allocation algorithm. Environment Explorer is a quasi-dynamic model in that it allocates land use and updates cell suitability in 1-year time steps, whereas Land Use Scanner is a comparative static model that allocates land use at one predetermined date at a time, or during a time period. Land Use Scanner has the potential advantage that the balancing factors could be interpreted, in terms of economic theory, as costs and benefits or even prices for land. In addition, the same kind of utility functions used in the sectoral models that generate the demand for land use could also be used in Land Use Scanner. Land Use Scanner is in this sense more open and extendable, which may offer some advantages. However, the actual implementation does not seem to acknowledge that utility models for different land uses and situations for choice do not, in general, yield comparably scaled utilities. Although Land Use Scanner does provide a unified theoretical basis for incorporating knowledge from experts or findings from other models, this assumes these can be suitably expressed in the same units. This potential problem does not occur in Environment Explorer because the algorithm performing the sequential assignment of land use remains consistent, if not a little arbitrary, in that no balancing of land uses is invoked.

In Environment Explorer, a random component is incorporated to account for a diversity of land users or non-explained behaviour. For presentation purposes several model runs are aggregated before presentation. Land Use Scanner does not incorporate a random component. Measures of goodness of fit for both models are high but this should not be interpreted as proof that the models are able to make reliable micro-level predictions. Rather these measures largely reflect the built-in inertia of the models, which maintain existing land uses, and as such they are heavily constrained to meet realistic land-use targets. This means that spatial patterns of land use changes will highly resemble those of past changes. When the goodness of fit of only the changes in land use is measured, the results are less impressive. In addition to the method of cell-by-cell comparison used, several other methods exist that measure different aspects of goodness of fit.

Both models assume that the suitability of cells remains constant within a round of assignments. Theoretically, however, assigning a land use to a particular cell immediately changes the suitability of that cell. Although time steps differ for Land Use Scanner and Environment Explorer, this means that suitability values for land use types depending upon other types, are not fully representative. Whether this is a problem in practice depends on how much real time is represented by one iteration of the model. There are examples in the literature of more detailed allocation algorithms.

Overall, the audit committee considers that Land Use Scanner and Environment Explorer suit their purpose and PBL's mission, i.e. assessing the likely impact of changing demand on aggregate land-use patterns in broad terms, and meeting the standards of academic quality and the state of the art in practice. The choice between the two models then becomes largely a matter of personal preference. One key question in this context is whether one wishes to use Environment Explorer for modelling regional and national processes or use more specialised and detailed exogenous models. In principle, either model could be used in that way, but Land Use Scanner has the potential advantage, at least in principle, of allowing the use of the same utility function that is also used in the model that provides the demand forecasts. It should be noted, however, that although this serial linking of models seems plausible, many experiences in other domains suggest that it may be more problematic than one would like to believe. This is especially true if feedback between models is important – the functions used in the different models may be based on inconsistent assumptions. It is therefore important to consider the effects of model chaining and the proliferation of errors in the kinds of model structures that are being considered. This problem becomes particularly acute if, for example, the models are nested in a series of higher-level models to obtain integrated assessments.

The audit committee was told that, increasingly, PBL is being confronted with policy issues that concern regional and even local scales and that, moreover, questions on policy issues were accompanied by related questions about management. In general, this means that the models need to be sensitive to a wider spectrum of policies and that other variables may become important. The committee considers that the current versions of Land Use Scanner and Environment Explorer are less appropriate for addressing policy issues below a national scale. The reason for this is that the drivers of change in both models are land-use cells and not the actual decision-makers (households, firms, developers, etc.), who constitute the truly dynamic elements in the system. Many changes with considerable environmental impacts (e.g. emissions) may take place in a city without much change in land-use patterns. In fact, land use shows substantial inertia and is one of the slower processes in urban dynamics. The prediction and assessment of such smaller-scale impacts would require simulating location choices and the space-time behaviour of households, firms and developers, along with the resulting changes in the spatial distribution of people and firms and, eventually, land-use patterns. A more behavioural orientation of the models would also require the introduction of demographic and economic variables, such as socio-economic groups, land and housing prices and transport costs. Environment Explorer does have a link to a transport model, but to date this is only an aggregate model of transport demand. For environmental issues such as exposure to air pollution and total energy consumption, a high resolution (in time and space) activity-based model of demand for travel and goods transport seems necessary to generate more reliable outcomes at the local level, which some policies and European guidelines require. Linking together several different models to incorporate these additional mechanisms is likely to result in these missing components only being addressed in superficial terms. To truly embrace the perspective of socio-economic activity, substantial model redesign or the development of a new model is required.

In principle, both models could be further elaborated along these lines. However, because Environment Explorer is already much closer to being an integrated model than Land Use Scanner, it seems most efficient to try and expand it further and examine whether it would be sufficiently sensitive to meet relevant policy requirements. This would involve adding a layer representing household and firm behaviour. However, in the long run it may be more effective to develop a completely different modelling approach that focuses on actors (people and firms) rather than on units of land use. Most integrated land-use transport models simulate the behaviour of households and firms, but fully operational models with the same spatial resolution as Land Use Scanner and Environment Explorer are still in their infancy. High-resolution agent-based models that include an activity-based model of transport demand, and which keep track of what activities are conducted where and for how long, constitute the frontier of academic work in this field and so far only partial results have been obtained. PBL has already experienced some of the considerable risks involved in following this approach.

Recommendations

In the *short term*, for the assessment of policies at a national scale, there is no major reason to change the current use of Land Use Scanner or Environment Explorer. The models may assume slightly different roles, with the former being used in combination with national demographic, housing and other models, and the latter being used as a desktop planning system. Some minor issues could be worked out in more detail – perhaps a task for Masters or PhD students. These issues involve making assumptions about socio-economic and political trends more explicit or making the models sensitive to a wider range of policies. Combining the two models does not deliver any advantages. The two models could, however, be used in parallel, with similar data sets, to explore the same policy options: significant differences in their outputs should be carefully probed for unexpected explanations.

In the *medium term*, PBL should consider developing a model that would be more appropriate for assessing policy impacts at regional and local scales. The audit committee recommends investigating the possibility of extending Environment Explorer's capabilities by adding a layer representing people and firms, introducing economics into the model and linking it to an activity-based model of transport demand. This would introduce the idea of agents into such a model, although it would be aggregated agents associated with population and employment at the cell level rather than individual agents.

In the *longer term*, a fully-fledged, integrated agent-based simulation framework may be desirable, especially if a more detailed approach is required to support regional and perhaps local policy-making. Such as model or framework will be fairly complex. Rather than starting from scratch, a definition study should explore which elements can be kept from the current models and how various existing micro-simulation models of demand for travel and goods transport, housing, firm demographics, use of green space, shopping behaviour, leisure activities, office development, agriculture, real estate, etc., may be incorporated. Many such models have recently been developed in the Netherlands and elsewhere and there is now a good opportunity to focus these efforts. In developing such a framework, constraints of data availability, calibration methods and computing time should be carefully considered. This may lead to a problem-specific, multi-level combination of aggregate and disaggregate model components, subject to the caveats regarding linked models (omissions, inconsistencies) already expressed above.

At the other extreme of complexity, PBL should explore the potential utility of much simpler, conceptual models, often called sketch planning models. Developing such models may serve some purposes better than detailed models, especially where broad, uncertain trends are concerned. These models would constitute needed links between detailed quantitative models, on the one hand, and the broad qualitative scenarios being developed by PBL, on the other. Large ensembles of such (meta-)models could be systematically generated, leading to a more scientifically grounded development of qualitative scenarios.

PBL should examine the ways in which different models can best serve the process of integrated assessment – both individually or in combination. This would involve a review of model chaining, of how models at different spatial scales might be interfaced, and how processes operating at different spatial scales can be successfully represented within integrated models.

This is a process that PBL is already engaged in and it involves continued vigilance with respect to how these models are integrated with those of other activities of PBL, as well as the models of other relevant Dutch agencies and research groups.

2.2.2 Resources

Are the resources available to PBL adequate and appropriate for the sort of use it makes of land-use models?

The audit committee believes that PBL basically has the right combination of people and skills to be able to liaise effectively with decision-makers. However, because different professional and public groups are interested in different aspects of modelling output, the sorts of decision-makers for whom PBL's models may be relevant should be further examined. Indeed, PBL's whole interface with clients should be examined, as mentioned below under Communication (Section 2.2.4).

The current research group has the right scientific balance and seems motivated to incrementally improve their tools and the process of communication, although an expansion of PBL's work into activity-based models, as recommended in Section 2.2.1, would require a strengthening of the expertise in that field. Staff are up to date with respect to new international developments in their field and national and international cooperation is high on their agenda. Furthermore, the national and international reputation of PBL has substantially increased of late.

PBL depends on others for developing LUMOS tools. There is a difference in this regard between the two models. Land Use Scanner has been developed in an open fashion, as part of a broader consortium following open source practices. It is therefore easier for PBL to influence its design. In contrast, Environment Explorer depends in part on proprietary software that is also being developed for other groups. While cooperation with external entities is both necessary and desirable, it is essential that PBL maintains the in-house expertise needed to conduct model analyses and make minor improvements to the models as needed. Eventually, a small number of PBL staff should have the expertise to develop new models jointly with consultants, as opposed to using models developed by others.

The LUMOS initiative is to be applauded and could be extended beyond harmonising the input of models, especially if PBL decides to develop agent-based models. It may even be worthwhile to consider LUMOS as a vehicle for enabling groups outside PBL to display and compare their models and to use the LUMOS website (www.lumos.info) as a forum for increasing national and international cooperation.

Because land-use modelling had recently regained momentum within the international research community, it is critically important that PBL staff remain at the cutting edge of the field. Joint projects and exchange programmes with universities (nationally and internationally), greater participation in international conferences and increased interaction with other agencies are examples of measures that should be considered for maintaining a high level of skills and expertise among staff. In addition, PBL should start positioning its modelling efforts in the perspective of the emergent information infrastructure, in particular the spatial data infrastructure and related networks of professional scientific activities that are interfacing with policy-makers and government. Such activities would also increase PBL's visibility as a highly desirable employer, thus attracting and retaining young highly-qualified scientists.

Recommendations

PBL should continue to develop collaborations in model development and data collection with other agencies and research institutions, but at the same time it should aim at maintaining a self-contained in-house capacity to conduct expert model analyses and make minor improvements to models without having to rely on external cooperation. In the longer term, some PBL staff should have levels of expertise needed to develop new models jointly with consultants, as opposed to always using models developed by others.

PBL should develop an infrastructure – at both national and international levels – for model development and dissemination built around the LUMOS initiative that interfaces with other types of models, research orientations, and data resources.

PBL should actively develop a programme for knowledge management and intensify national and international collaborations in order to constantly update the skills and expertise of its staff. This would help make PBL less dependent on external consultants and would further enhance its national and international reputation.

2.2.3 Use of Results

Are the results of studies using land-use models adequately used in policy report making processes inside PBL and in joint projects with other, relevant national and international agencies?

Land Use Scanner and Environment Explorer – and, indeed, PBL itself – owe their existence to the contributions they makes to policy-oriented decision-making. The appropriateness and scientific quality of the models *results* are therefore every bit as important as those of the models themselves. The committee was informed about several applied projects, in particular those focusing on scenario development. This is a very important activity as it combines quantitative with qualitative analysis, and links the two land-use models to issues in both policy formulation and to external developments beyond the purview of policy-makers. As staff have indicated, these projects have provided important learning experiences and there seem to be ample opportunities to further improve the process leading to such applications.

The audit committee did not review all the applications that had been carried out by PBL using the two models or the ways in which their results were communicated to policy-makers. However the committee was concerned to find a lack of a clear distinction between the systematic development and examination of consistent policies and policy packages, on the one hand, and the generation and exploration of future scenarios, on the other. The former concern courses of action that policy-makers may adopt, whereas the latter outline potential developments in the environment and society at large that policy-makers have little or no influence, over: megatrends as individualisation and globalisation having an impact on the demand for land. The distinction between policies and scenarios is important because it is difficult, if not impossible, to distinguish the contribution to the outcomes of individual assumptions or policies from the results of scenario simulations: this is only possible if individual policies or combinations of policies (policy packages) are formulated and examined separately from the scenarios themselves. This distinction is essential to the process of decision and planning support and should provide the essential framework within which to set up appropriate applications.

The distinction between policies and scenarios would also be helpful in further articulating the role of uncertainty in applied projects. Currently, uncertainty appears not to receive the systematic attention it deserves. Several aspects of uncertainty are relevant. First, there is the uncertainty of global and local events not under the control of decision-makers that are likely to have an influence on future land-use patterns (megatrends). Second, there is the uncertainty about the responses of the system of interest to policy instruments. Third, land-use decisions of particular actors are made under conditions of inherent uncertainty regarding the decisions of others, including spatial planners. Fourth, there is uncertainty about the value choices that future stakeholders will make. Finally, there is uncertainty about data (for example accuracy, representativeness) and model predictions. The generation of both future scenarios and policy packages should take these different kinds of uncertainty systematically into account.

This uncertainty is also related to the calibration of the models, especially when used in scenario-based studies. Currently model calibration is the result of a combination of expert knowledge and statistical analysis. Statistical analysis is, of course, very important and successful reproduction of changes in land-use patterns provides some evidence of the validity of the models. However, it is equally important that any assumptions made are shared by experts outside PBL and have institutional (PBL) legitimacy. In addition, validation should also rely on meta-analysis of assumptions about individual behaviour and not only on process outcomes.

Finally, it seems as if many of the cases in which models are currently used are focussed on impact assessment. This seems to be unnecessarily narrow in scope as the data, models and expertise could also be used to support the policy-making process, including generating plans, formulating instruments and developing standards and guidelines. Concepts such as robust adaptive planning are of potential value in this context since they would help connect models, policies and scenarios within a more systematic framework.

Recommendations

PBL should further expand the methodology for developing scenarios for policy-making by systematically differentiating between exogenous developments and individual policies and policy packages. In this context, exploration of the role of formal decision-making and planning support systems should continue.

The issue of uncertainty deserves further attention in relation to model development, model application and scenario development and assessment. Qualitative, as well as quantitative, notions of uncertainty should be associated with the various model structures in terms of their inputs, outputs and mechanisms.

Uncertainties of different kinds also lie at the core of scenarios of the future and should, therefore, be systematically considered both within individual scenarios and in connection with ensembles of scenarios characterised by different degrees of plausibility.

PBL should consider developing a wider range of qualitative and quantitative decision-support tools to be able to expand its role beyond impact assessment. In particular, the methodology of robust, adaptive planning has the potential to integrate models, policy options and scenarios of exogenous or autonomous developments within the same systematic framework. As part of such a framework, certain kinds of simpler conceptual models may be developed to help fill gaps between detailed quantitative models, on the one hand, and qualitative future scenarios, on the other (see recommendations in Section 2.2.1).

2.2.4 Communication

Are the results of studies using land-use models reported in a way that communication with policy-makers is well targeted and optimised, while maintaining an acceptable level of scientific quality?

Insights gained from models can be communicated to policy-makers in a variety of ways. First, model software can be used in workshops and policy-development sessions to provide participants with a better understanding of the issues, underlying processes and likely effectiveness and impact of alternative policy instruments. Environment Explorer especially stands out in this regard, as it has a very user-friendly user interface that does not require any knowledge of scripting and allows for the exploration of policy strategies at every level.

Second, outcomes of the models can be communicated to policy makers. Visualisation will clearly play an important role in this regard, although there is a risk that visualisation may become a goal in itself and distract attention away from key messages. Given the problem of stochastic variation at the micro level as discussed earlier, care should be taken to match any visualisation with the spatial resolution at which reliable results can be achieved. In general, this will preclude the presentation of results at the micro-level, except for demonstration purposes, that is, to focus attention on the essential information. While current efforts are very promising, the audit committee considers that there still is much work to be done on developing visualisation methods that are effective and suitable for policy applications. Both models clearly enable good visualisations to be developed, including animations, especially in the case of Environment Explorer. 3-D visualisation shows promise in that it helps engage non-expert decision-makers and other stakeholders in scenario testing and development.

In addition to communication with policy-makers and other stakeholders, communication with scientists remains of utmost importance. If the models used by PBL do not meet international standards, the credibility of their applications may become an issue. Regular publication of scientific articles in peer-reviewed journals is important in this respect.

Recommendations

Current visualisation tools should be further developed and expanded upon without losing sight of the ultimate goal: communicating essential information.

PBL should explore different ways of using visualisation to communicate the results of scenario development and policy studies simulated by the models.

Just as publishing scientific articles on modelling research is desirable, publishing in appropriate media about improvements to, and novel applications of, visualisation techniques in the context of policy exploration should be encouraged.

2.2.5 Next Steps: The Coming Years

Can the auditors advise PBL on its data, model and application methods for the coming years?

The recommendations listed in the previous sections reflect differing degrees of scientific urgency and practicality. Given a thorough knowledge of its own resources and limitations, PBL is in a better position than the audit committee to determine the most appropriate timing and balance of the efforts that need to be made to meet the audit committee's recommendations. Considering that these were formulated to be as consistent as possible with PBL's current mode of operation and resources, the committee believes that all recommendations could be met within a 5-year time frame. Thus, we strongly recommend the development of a detailed 5-year plan to that effect.

The committee considers that some minor issues related to Land Use Scanner and Environment Explorer can be dealt with without a need for much planning. Moreover, strategically, it is important to start identifying early on (a) the new kinds of policy questions that are being asked, (b) how the socio-economic dynamics can be captured in a new, additional model or model framework, (c) what additional data would be required for these, and (d) to what extent could the effort to develop a new model benefit from the (expanded) LUMOS consortium. In this, the role of PBL should also be identified and the bureau positioned with respect to other agencies, such as DVS (Road and Water Transport Service), because these will be dealing with similar issues in the near future.

Recommendations

The audit committee considers that PBL should develop a phased plan to implement these recommendations over the next 5 years, by, say, 2012. Our proposals are quite consistent with the working style of the current research group and PBL as a whole and have been made on the basis that these can be realised with relatively little change to the organisational structure of the group within PBL.

Using the LUMOS consortium as a platform, PBL should reach out to agencies inside and outside the Netherlands to promote Land Use Scanner and Environment Explorer and educate others about their use. This could include engagement in some limited training of other professionals and some technology transfer of the models.

2.3 Conclusion

This chapter reports the findings of an audit conducted to evaluate the scientific quality of the processes and products of PBL relating to the application of land-use models primarily at the national level. The audit committee thinks that the Land Use Scanner and Environment Explorer are among the best in the field, and are very useful for assessing policy impacts relating to the spatial distribution of land use at the national level. There are several technical elements that could be further developed but this will not dramatically change the adequacy of these models for their intended purpose.

PBL is however also increasingly confronted with questions about the regional and even the local scale and questions about management. In general, this means that the models should be sensitive to a wider spectrum of policies, and that other variables may become important. To address such questions and truly embrace the socio-economic activity perspective that is more adequate for such questions, substantial model redesign or the development of a new (activity-based) model is required.

In addition, the Agency may consider simpler models for policy development as opposed to policy assessment. Intensifying collaboration with other agencies in the Netherlands and elsewhere in terms of promoting the models and educating other agencies and groups on their use, building on the LUMOS consortium, should be the core of an improved communication strategy.

Reference

Timmermans, H., Batty, M., Couclelis, H., & Wegener, M. (2007). *Scientific audit of national land use models; Report and recommendations of the audit committee*. Bilthoven: Netherlands Environmental Assessment Agency (MNP).

Chapter 3
Core Principles and Concepts in Land-Use Modelling: A Literature Review

Jonas van Schrojenstein Lantman, Peter H. Verburg, Arnold Bregt, and Stan Geertman

3.1 Introduction

Simulation models of land use predict or describe land-use change over space and time. Recent overviews of land-use simulation models show an overwhelming amount of different types of models and applications (Heistermann, Muller & Ronneberger, 2006; Koomen, Stillwell, Bakema & Scholten, 2007; Verburg, Schot, Dijst & Veldkamp, 2004). Obviously, such models are simplifications of reality, but increasing computing power over the years has made it possible to incorporate more and more complexity in such models. This increased complexity, however, tends to obscure the theoretical foundations of land-use simulation models. This theoretical foundation relates to the core principles that are used to explain land-use change and the concepts that are applied to translate these principles into a functioning model of land-use change. An in-depth review of land-use change concepts, their underlying principles, applicability and translation into actual models does not exist to our knowledge. In this chapter we aim, therefore, to analyse the application of various theoretical concepts of land-use change that are used in modelling. This analysis is a first step to better understand the conceptual background of land-use change and the application of these concepts in computer simulation models. Based on this review we present some observations on important research issues in land-use modelling and suggest possible ways for further model improvement.

Computer simulation models of land use are characterised in terms of core principles and applied concepts in Section 3.2. After this discussion of the theoretical background of current land-use models we dwell on two additional issues that are important for improving simulation models of land-use change. Lessons from model validation efforts are described in Section 3.3, while the importance of scale issues in land-use change is dealt with in Section 3.4. The main conclusion and discussion can be found in Section 3.5, whereas some themes for future research are suggested in Section 3.6. The latter suggestions are specifically focused on the Dutch LUMOS models.

J. van Schrojenstein Lantman (✉)
Nelen & Schuurmans, PO Box 1219, 3500 BE Utrecht, The Netherlands
e-mail: jonas.vanschrojenstein@nelen-schuurmans.nl

E. Koomen, J. Borsboom-van Beurden (eds.), *Land-Use Modelling in Planning Practice*, The GeoJournal Library 101, DOI 10.1007/978-94-007-1822-7_3,

3.2 Categorising Land-Use Simulation Models

In the literature, various categorisations of computer simulation models of land-use change have been made. Baker (1989) categorised them according to scale: whole landscape models, distributional landscape models and spatial landscape models. Briassoulis (2000) categorised models according to the modelling tradition to which they belong: statistical/econometric, spatial interaction, optimisation, integrated and 'other' modelling approaches. Lambin, Rounsevell, and Geist (2000) distinguished between empirical-statistical, stochastic, optimisation, dynamic simulation, and integrated modelling approaches. Agarwal, Green, Grove, Evans, and Schweik (2001) analysed 19 different computer simulation models according to a three-dimensional framework: space, time, and human decision-making. Verburg et al. (2004) discussed computer simulation models according to six features: level of analysis, cross-scale dynamics, driving factors, spatial interaction and neighbourhood effects, temporal dynamics, and level of integration. Heistermann et al. (2006) classified 18 computer simulation models according to geographical, economic and the integration of both. Koomen and Stillwell (2007) did not categorise models at all, but rather discussed a number of characteristics of computer simulation models: static/dynamic, transformation/allocation, deterministic/probabilistic, sectoral/integral, zones/grid.

Common factors in these categories of computer simulation models can be recognised (for example, distinction between dynamic and integrated models), but differences in approach make these studies difficult to compare. The general complexity of computer simulation models and the fact that the field of land-use modelling is interdisciplinary are the most probable reasons for the different categorisations made by these authors. Yet, despite the marked differences between computer simulation models of land-use change, they do have a common basis.

It is common practise for modellers to describe the processes of land-use change according to a particular concept or mechanism that can be used to characterise land-use changes (e.g. Cellular Automata). In fact, a concept of land-use change is composed of a set of core principles according to which 'real world' processes of land-use change are assumed to operate (e.g. Cellular Automata is based on neighbourhood interaction). This concept of land-use change is then codified into algorithms, for which other, different algorithms can be used, which will in turn lead to different computer simulation models (e.g. Environment Explorer or GEOMOD2). Thus, a land-use change algorithm is no more than the translation of a concept of land-use change into calculation rules for a computer simulation model.

Figure 3.1 shows the relationship between the process of land-use change and computer simulation models. The grey box encloses the steps that are the focus of this chapter. Although different divisions can be made, all computer simulation models have in common that they are based on four core principles. The main steps in Fig. 3.1 are discussed in the following subsections.

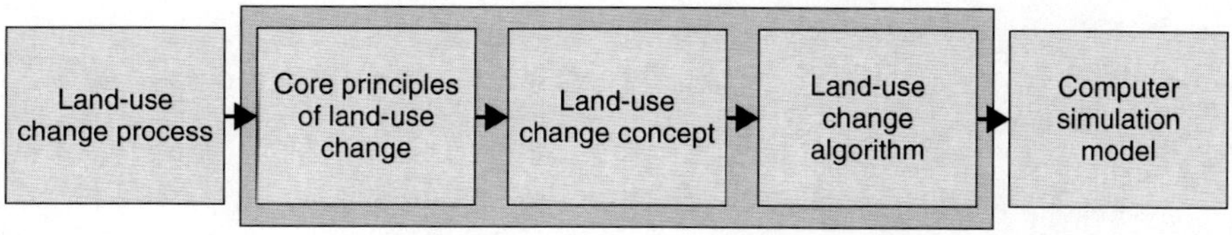

Fig. 3.1 Relationship between the land-use change process and computer simulation models

3.2.1 Land-Use Change Process

'Land use change is the result of socio-economic and biophysical phenomena, dependent on spatial location, scale, and existing land use' (Briassoulis, 2000; Lambin et al., 2001; Lesschen, Verburg & Staal, 2005; Meyer & Turner, 1992; Turner et al., 1995). In our literature review we focused on concepts of land-use change and their origin, not on the process of land-use change itself. Nevertheless, the literature mentioned in this subsection provides a good starting point for reading on land-use change processes.

3.2.2 Core Principles of Land-Use Change

In-depth analysis of the literature on land-use modelling shows that all simulation models of land use are based on at least one of the following four core principles of land-use change:

1. Continuation of historical development;
2. Suitability of land (in monetary or other units);
3. Result of neighbourhood interaction; or
4. Result of actor interaction.

Continuation of Historical Development

The premise behind historical land-use change is that future land use can be predicted by means of historical changes. This can be interpreted in several ways: in the past people liked to live near the water, so in the future they will like to do so too; or 15% of the forest became residential land so it is likely that given the same time horizon this will happen again.

Extrapolating land-use change from past changes into future changes is an intuitive concept and is, therefore, widely used. A business-as-usual scenario is a practical application of this principle, as demonstrated by Kuijpers-Linde et al. (2007).

Suitability of Land (in Monetary or Other Units)

Suitability may cover a combination of factors, such as soil suitability, spatial location or terrain characteristics of a piece of land; distance to a market is an aspect of spatial location. The underlying premise is that people want to maximize profit, which can be expressed either in monetary units (quantitative) or non-measurable units (qualitative).

Result of Neighbourhood Interaction

The principle of neighbourhood interaction states that the possibility of transition from one use of land to another is dependent on the land use of its surrounding cells. This driver can be biophysical, e.g. a land cover affecting that of neighbouring cells, or a socio-economic one; the latter can be explained with, for example, the Core and Periphery model, which assumes that people want to maximise profit (Fujita, Krugman & Venables, 1999; Krugman, 1991; 1999).

Result of Actor Interaction

The main assumption in actor-focused decision-making is that land-use change is the result of interaction between actors. Actors can be represented as agents: a single entity or a group of actors, depending on the scale of modelling. The core principle of actor decision-making aims at explaining and understanding socio-economic drivers and policies for development. A complicating factor is that it is only recently that surveys have been held among actors in certain case studies in which their preferences were monitored. Therefore, the amount of data is insufficient and it is difficult to validate them. In practice this core principle of land-use change modelling is still in its infancy, but it seems to be a promising research tool (Matthews, Gilbert, Roach, Polhill & Gotts, 2007).

3.2.3 Land-Use Change Concepts

Different concepts of land-use change represent differing attempts to enable science to explain and thus translate reality on the ground into a model; each concept brings with it its own advantages and disadvantages. Any concept of land-use change is always based on one or more of the four core principles described in the previous subsection.

Quantity of Subjects in the Literature on Land-Use Change Modelling

The following well-known concepts of land-use change were selected from the literature:

- logit functions;
- markov chains;
- cellular automata (CA); and
- agent-based modelling.

A search of the Scopus database was made for each concept. With access to over 15,000 peer-reviewed journals (Scopus, 2008), Scopus is the largest abstract and citation database of research literature and quality web sources. In addition to basic information such as author(s), year of publication and name of journal, the citation count for each concept was compiled. As references in Scopus only go back as far as 1996, articles published before then were not taken into account.

Subject experts selected the following search terms:

- 'artificial neural networks' AND 'land use';
- 'cellular automata' AND 'land use';
- 'agent-based' AND 'land use';
- 'agent based' AND 'land use';
- 'Markov' AND 'land use'; and
- 'logit' AND 'land use'.

The search was performed in February 2008 and the papers found were classified according to the concept in use. Papers in which two different concepts are compared were counted as an occurrence of both concepts. A paper describing a model combining two (or more) concepts was counted as hybrid paper. Some papers described concepts only in theory, while other papers described concepts as applied in case studies. After the analysis, the database was checked for any key papers that were missing; these were added when necessary. Figure 3.2 shows the results of the literature search. It was remarkable that, with hybrid and CA papers combined, CA appeared in over 40% of the papers published on land-use modelling. Apart from the obvious fact that CA is a popular concept among scientists modelling land use, other explanations are possible. One of them is that when an article is written on the use of cellular automata, it receives this keyword

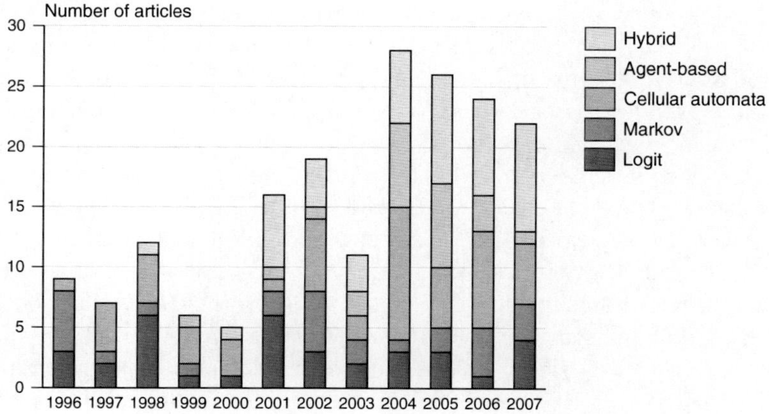

Fig. 3.2 Scientific articles in Scopus[1] on various concepts of land-use change

[1]Scopus is the largest abstract and citation database of research literature and quality web sources. It is designed to find information scientists need. Quick, easy and comprehensive, Scopus provides superior support for literature searches. Surf to www.scopus.com.

from the author. If, however, an article is written about a land-use model based on a logit concept, the keyword selected by the author may not be 'logit'. Our Scopus literature search is, therefore, merely an indication of the number of articles written about certain concepts of land-use change.

3.2.4 Concepts of Land-Use Change in Practice

The following concepts of land-use change were identified in the literature:

- cellular automata;
- statistical analysis;
- Markov chains;
- artificial neural networks;
- economics-based models; and
- agent-based systems.

In this subsection, each concept will be discussed in terms of the basic principle of the concept and its use in practice.

Cellular Automata (CA)

The most well-known land-use change concept is that of cellular automata (Langdon, 1998; White and Engelen, 1994). The basic principle of CA is that land-use change can be explained by the current state of a cell and changes in those of its neighbours. It is, thus, based on the core principles of continuation of historical development and result of neighbourhood interaction. In other words: if a road through a forest is paved, this enhanced the connectivity of the region, with the result that forest land use is replaced by residential use. CA comprises four elements: cell space; cell states; time steps and transition rules (White & Engelen, 2000). Decision rules can be expert-based or derived from statistical analysis. Two main types of CA can be distinguished: unconstrained and constrained. Unconstrained is the most 'true' CA, as it only uses decision rules to calculate land-use change. In constrained CA, the amount of land-use change per land use class is limited; the limit of a certain land-use class is either expert-based or calculated from historical land use.

Made famous by Gardner (1970), John Conway's game of life is the best known example of a cellular automaton to date. Tobler (1979) was the first to introduce the use of cellular automata in geography. In the decades that followed this was further developed by Couclelis (1985), Batty and Xie (1994), and White and Engelen (2000). More recently, Hagoort (2006) has put together a nice overview of the history of CA, and Pinto and Antunes (2007) provide an overview of the use of CA in urban studies. In the Netherlands, a constrained CA model (Engelen, White, Uljee & Drazan, 1995; Geertman, Hagoort & Ottens, 2007) called the Environment Explorer (part of the LUMOS toolbox), is commonly used in spatial planning and policy-making in (de Nijs, de Niet & Crommentuijn, 2004).

Statistical Analysis

All kinds of statistical information can be derived from land-use maps. Statistical information can be based on all four core principles, depending on the aim of the research. A widely used application is the computer program FRAGSTATS by McGarigal and Marks (1995), which can analyse ample amounts of landscape statistics. In the discipline of land-use modelling, linear regression, probit regression, binomial logit and multinomial logit models are used to analyse statistical relations in land use and predict future patterns. Logit analysis provides an interpretable linear model (deMaris, 1992). Logistic regression is used to analyse the probability of occurrence, dependent on different factors, of a certain land-use category, for instance biophysical characteristics, distance from location to the market (Verburg, Ritsema, van Eck, de Nijs & Schot, 2004). The coefficients for the conversions can be calculated from historical land use and extrapolated to the future. A logit model can be based on neighbourhood interaction, historical land-use change, soil suitability, spatial location or combinations of these. A binary logit model is a mathematical variant of an ordinary linear model that will give a prediction of land-use change relative to all other options. A multinomial logit (MNL) model is similar, with the exception that it describes the conversion of different land-use categories relative to a reference category (deMaris, 1992; Liao, 1994; Wrigley, 1976).

Among the first to apply logit models in relation to land-use change were Veldkamp and Fresco (1996), with the CLUE model, and Wear and Bolstad (1998), who linked a negative binomial regression model of building density with a logit model of land use. An example of MNL analysis is given in Walsh, Soranno, and Rutledge (2003), where spatial association of various categories of land use is analysed. Land Use Scanner (Hilferink & Rietveld, 1999), which has been used in producing sustainability outlooks for the Netherlands (Borsboom-van Beurden et al., 2005; Kuijpers-Linde et al., 2007), is an example of the use of logit modelling in the planning and policy environment.

Markov Chains

Burnham (1973) was one of the first to propose using Markov chain analyses for modelling land-use change. Such analyses are based on the core principle of a continuation of historical development. A Markovian analysis uses matrices (e.g. Table 3.1) to represent changes between land-use categories.

Table 3.1 Example of a Markov matrix

Current\future land use	Farmland	Peri-Urban	Urban
Farmland	0.50	0.40	0.10
Peri-urban	0	0.8	0.2
Urban	0	0.1	0.9

Under an assumption of stationarity (temporal rate of change and amount of change stay the same), the matrix can be used to calculate the probability of land-use change of one land-use category to another. A disadvantage of this type of analysis is that it is non-spatial, meaning that additional assumptions are required for allocation. Distinction is made between a first-order and second-order Markov matrix. The former uses a matrix with current land use and a change matrix based on expert knowledge, while the latter calculates changes from one land use to another by comparing two maps of land use over time, that is, this change matrix is constructed from historical land-use change. The temporal rate of change is assumed to be constant, so the change matrix can then be used in the same way as in first-order Markov analysis.

Examples of Markov chains in land-use studies are given by Muller and Middleton (1994), Fearnside (1996) and Lopez, Bocco, Mendoza, and Duhau (2001). Because of its simplicity, Markovian analysis is a popular technique to combine with other concepts. The GEOMOD2 model (Pontius, Cornell & Hall, 2001) is an example of combining Markov techniques with statistical analysis.

Artificial Neural Networks

The use of artificial neural networks (ANNs) in land-use modelling has increased substantially over the last few years because of advances in computing performance and the increased availability of powerful and flexible software (Skapura, 1996). ANNs are self-learning computer models and are used for pattern recognition in many disciplines (Pijanowski, Brown, Shellito & Manik, 2002). Recently, ANNs have found their way into the world of land-use modelling. The ANN algorithm assumes a relation between past and future land-use change and can be linked to suitability maps. ANNs can be based on all four core principles. The first to apply ANNs to a computer simulation model was Pijanowski et al. (2002). The model 'trains' itself on a dataset and the corresponding land-use maps of different years enabling it to recognise and reproduce the pattern of land-use categories (Mas, Puig, Palacio & Sosa-Lopez, 2004; Pijanowski, Pithadia, Shellito & Alexandridis, 2005).

Economics-Based Models

Although not exactly a concept of land-use change, but rather more a land-use theory, economics-based models of land use cannot be left out of the list of concepts of land-use change. These models are based on the core principle of suitability of land (in monetary or other units), although the core principle of continuation of historical development can also be included. The first theory on land use was developed by Johan Heinrich von Thünen in 1826. It was translated into English in 1966 (von Thünen, 1966). Von Thünen converted the principle of Ricardo (1817) that profit will be used to reinvest in a land-use change concept. Von Thünen stated that as long as the profit of a commodity (turnover minus production costs) is higher than the transportation costs of the commodity, the land will continue to be used to produce this commodity. Alonso (1964) expanded this theory by including land-use suitability and an individual bid-price curve for every household or firm. A little later Sinclair (1967) expanded von Thünen's theory to explain urban sprawl. More recent applications of the theory have been reported by Chomitz and Gray (1996), Nelson and Hellerstein (1997), and Walker (2004) which all use von Thünen's model as the base theory to explain tropical deforestation.

Agent-Based Systems

Be it agent-based modelling, individual-based modelling, micro-simulation or activity-based modelling, the common ground of all these concepts of land-use change is that they are based on the core principle of actor interaction. The differences between these approaches are to be found in the institutional dimension of scale, for example, ranging from individual modelling to group behaviour and the number of agents being modelled. One of the first actor-interaction based models was a competition model for individual trees (Bella, 1971), but it took a long time before this type of model was applied to land-use change. Balmann (1996) was one of the first to introduce a spatially dynamic model of land-use change for analysing structural change in agriculture in which individual farmer preferences were taken into account. Since 1996 the number of articles on agent-based land-use models has steadily grown over the years (Fig. 3.2). Within agent-based systems, we distinguish four sub-types: agent-based modelling; individual-based modelling and micro-simulation; activity-based modelling; and expert-based decision rules.

An agent-based model of land-use change consists of two key components: a map of the study area, and a model with agents that represent human decision-making (Parker, Berger & Manson, 2001). The preferences of agents can be defined by expert judgement, by using questionnaires, or by using an artificial neural networks technique. A multi-agent system is a set of agents interacting in a common environment, which consists of everything in the model that is not an agent (Ferrand, 1996; Le Page, Bousquet, Bakam, Bah & Baron, 2000). An agent is a representation of an actor in the process of land-use change and can be either an individual or a group. The following actor interactions can be distinguished (Ligtenberg, Bregt & Lammeren, 2001):

- interactions between the spatial objects of the environment and actors;
- interactions between spatial objects;
- interactions between actors and spatial objects; and
- interactions between actors and other actors.

Originating in the discipline of ecology, *individual-based modelling* is a subset of agent-based modelling. It differs from the agent-based modelling usually done in that the scale of the concept is always at the individual level. Each individual is represented as an agent, each with their own preferences (Grimm et al., 2006). In social sciences, micro-simulation models aim at reproducing human behaviour at the individual level (Moeckel, Schurmann & Wegener, 2002). Micro-simulation is mostly used to model urban land use and transport that is under development; an example of this approach is the Ilumass project (Wagner & Wegener, 2007).

Activity-based modelling is a special subset of the agent-interaction decision-making concept. A relatively new concept in research on land-use modelling, activity-based modelling is based on the work of Arentze and Timmermans (2000) and the UrbanSim model (Waddell, 2002). In the Netherlands, there is need for policy-makers to know the impact of intensification of cities as compared to that of urban sprawl. In activity-based modelling, the relation between mobility and infrastructure and their impact on land-use change is explicitly addressed which in turn can be used to recognise, for example, traffic bottlenecks. The concept of activity-based modelling is relatively new and opinions on what an activity-based model exactly is differ. It can range from modelling of population density in combination with CA, to a sort of individual-based model in which, for example, the result of movements per individual during the day is taken into account, which can help recognise possible traffic jams.

In cases of uncertainty about parameters, *expert-based* rules, which encapsulate experts' knowledge of a specific area, are used to estimate them. Basically, all four core principles can be used for this. The IMAGE model (Alcamo, Kreileman, Krol & Zuidema, 1994a; 1994b) is an example of a land-use model built on expert-based decision rules.

Analysis of concepts of land-use modelling in practice and the models involved allows modellers to pick the most appropriate concept of land-use change for the study area under investigation. The relation between land-use change concepts and underlying core principles is shown in Table 3.2.

3.2.5 Selection of Operational Land-Use Simulation Models

This subsection contains a selection of computer simulation models of land-use change. Various models have been developed that can estimate future land use.

Table 3.2 Relation between concepts of land-use change and core principles

Concept of land-use change	Continuation of historical development	Suitability of land (in monetary or other units)	Result of neighbourhood interaction	Result of actor interaction
Economics-based models	P	A	–	–
Agent-based models	P	P	P	A
Cellular automata	A	P	A	–
Statistical analysis	P	P	P	P
Markov chains	A	–	–	–
Artificial neural networks	P	P	P	P
Individual based modelling and micro-simulation	P	P	P	A
Activity-based models	P	P	P	A
Expert-based decision rules	P	P	P	P

A = Always, P = Possible, – = Not possible or not practical

For reviews of some of these models see Agarwal et al. (2001), Hunt, Kriger, and Miller (2004), and Verburg et al. (2004). All such models use different concepts and algorithms to decide where and when land-use change will take place. It is beyond the scope of this chapter to describe all the land-use models available: only models from the LUMOS toolbox and a small selection found in the literature are described. The models are described only briefly here as computer simulation models of land-use change are not the focus of this chapter.

The LUMOS toolbox consists of Land Use Scanner and Environment Explorer (LUMOS, 2005). *Land Use Scanner* allocates land according to bid prices for various types of land use (Hilferink & Rietveld, 1999; Koomen, Loonen & Hilferink, 2008). The possibility of government intervention in determining land use is taken into account by, among other things, adding aggregate constraints. *Environment Explorer* is built on a local-level Cellular Automata model constrained by regional-level spatial interaction model (White & Engelen, 2000).

Other well-known land-use models are:

- *CLUE* (Conversion Land Use and its Effects) is a dynamic simulation model using empirically-derived relations between land-use change and driving forces from cross-sectional analysis at multiple scales (Veldkamp & Fresco, 1996; Verburg, de Koning, Kok, Veldkamp & Bouma, 1999; 2004).
- *Land Transformation Model (LTM)*, which combines GIS and ANNs to predict future land use. The ANNs are used to learn the patterns of development in a region, whereas GIS is used to develop the spatial-predictor drivers and perform spatial analysis on the result (Pijanowski et al., 2002).
- *GEOMOD2* can be used to model both forwards and backwards in time. It selects locations for a particular type of land use based on: nearest neighbourhood interaction; political sub-region; and the pattern of biophysical attributes (Pontius et al., 2001).
- *SLEUTH* (Slope, Land use, Exclusion, Urban extent, Transportation and Hill shade), formerly called the Clarke Cellular Automaton Urban Growth Model, was developed for and tested on various cities in North America. It is based on cellular automata and is designed to be scalable and universally applicable (Silva & Clarke, 2002).
- *UrbanSim* uses aggregate economic and spatial-interaction models, and pursues a disaggregated approach to predict changes over small time steps. It takes into account the demand for real estate at each location, and the actors and choice processes that influence patterns of urban development and real-estate prices (Waddell, 2002).
- *IMAGE* is an ecological–environmental framework that simulates the environmental consequences of human activities worldwide. The objective is to explore the long-term dynamics of global change. It has a relatively coarse resolution (of 5 min $\times$ 5 min) and the allocation of new land use is based on demand, location preferences and autonomous system change (Alcamo et al., 1994a; b).

- *CORMAS* stands for Common-pool resources and multi-agent systems. It is a multi-agent framework in which interactions between groups of agents and a shared environment with natural resources can be simulated (Le Page et al., 2000).
- *ILUMASS* is a micro-simulation model of urban land use. Unfortunately, it never became fully operational, but it is a good experiment on micro-simulation of urban land use. It takes into account Land Use (Population, Accessibility, Firms), Transport (Travel demand, dynamic traffic assignment, demand for goods transport) and Environment (impacts, emissions) (Wagner & Wegener, 2007).

3.2.6 Core Principles and Concepts in Practice

Our literature research has enabled us to compile an overview of the relationships between core principles, land-use change concepts, land-use change algorithms and land-use change models in practice; see Fig. 3.3. The figure links the core principles with the modelling concepts that are normally associated with it and, by way of example, indicates examples of existing land-use models that rely on that particular combination of principles and concepts. Since statistical analysis, artificial neural networks and agent-based systems can all include one of the core principles, they are connected to all the core principles. The same applies to the computer simulation

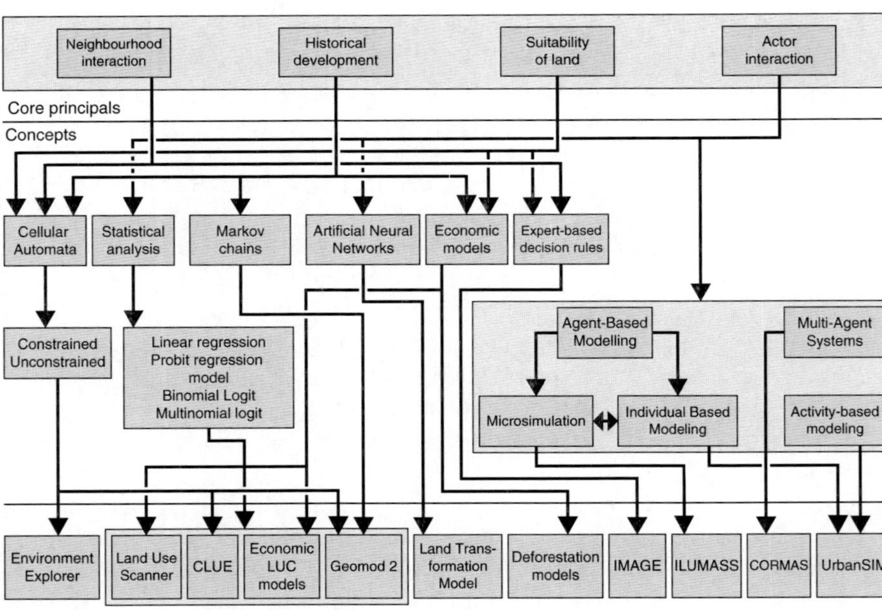

Fig. 3.3 Examples of relationships between core principles, concepts and land-use simulation models in practice

models of land-use change, for example, Land Use Scanner, CLUE, economic land-use change models and GEOMOD2, all of which incorporate statistical analysis.

This detailed overview has been compiled from the results of our literature research. Since models are mostly developed further after the initial paper is published, it is possible that some models now incorporate other or more core principles or land-use change concepts than is shown in Fig. 3.3; the same relationships between core principles and concepts of land-use change are also shown in Table 3.2.

3.3 Lessons from Model Validation

Validation is an essential step in assessing the performance of any land-use model (see, for example, Pontius et al., 2008). In this section we list some general lessons that can be learned from model validation and that are relevant for the further development of land-use simulation models.

Validation of land-use models consists of two elements: validation of the spatial resemblance of the output maps with reference maps; and validation of the resemblance of the described land-use change process. If a model is able to predict spatial resemblance accurately but fails to predict the process, it is questionable whether policy-makers can learn much from the results. If, on the other hand, the model is more process oriented but fails to predict spatial resemblance, it is also of little use.

In general it can be said that an inductive (data-driven) approach is strong in reproducing land-use patterns which is (spatial resemblance), but is weak in explaining correlations found. A deductive (theory-driven) approach, on the other hand, is strong in explaining how and why land use will change (which is process resemblance), but is weak in spatial allocation of land-use change (Overmars, de Groot & Huigen, 2007).

Overmars et al. (2007) have shown that prediction of spatial location with deductive models is promising when compared to inductive models. For land-use change developments that follow a single dominant process it is expected that a deductive model will score only marginally less on spatial resemblance than an inductive model (see Fig. 3.4). For land-use changes that result from multiple processes, the differences in spatial and process resemblance between inductive and deductive simulation models increases.

One could argue that inductive simulation models, in which a lot of correlations are found, are more suitable for allocating land use at the correct spatial location. The obtained correlations, however, do not provide a causal relation and that explains why land-use change occurred at specific locations. At the other end of the spectrum are deductive models, which are aimed at modelling the process behind land-use change. Although these theory-driven models approximate the governing processes more closely than inductive models, they have more difficulty in identifying the exact spatial location of change. But since these kinds of models

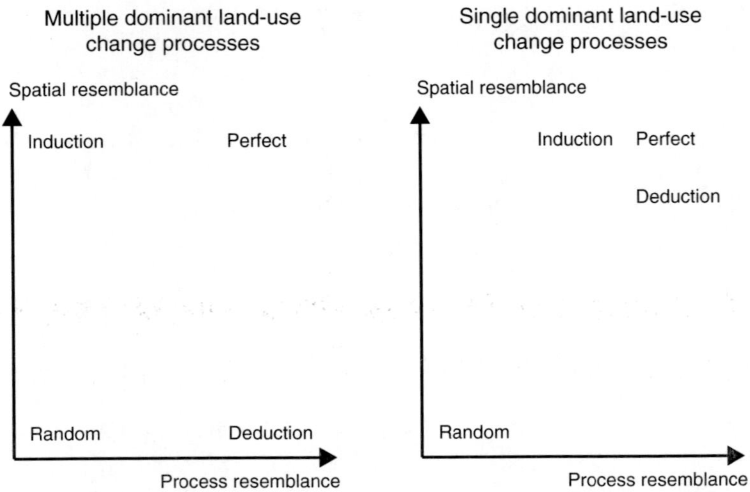

Fig. 3.4 Process resemblance versus spatial resemblance for various types of models in relation to the complexity of underlying land-use change processes

provide more insight into causality, they enhance the learning process more than inductive models.

In general, strong process resemblance is of more use to policy-makers than spatial resemblance because they can learn from the process and the interactions that take place. Ideally, however, a simulation model should be able to meet spatial and process resemblance criteria. To validate a model on its process accuracy is a difficult task since the process of land-use change is often very complex, so it is impossible to check whether the model accurately has reproduced the process. A possible technique for checking whether all key processes have been included in a model is to add other processes to the model. If the outcome, such as the land-use pattern, does not change significantly it indicates that all key processes were initially included in the model (or, at least, that the new process that was added to the model is not important in producing spatial patterns. Thus, the key processes can be identified by a combination of common sense and trial and error. To check the stability of the model, the same process can be modelled several times. If the outcomes are not comparable, it indicates that the model is not stable enough to use for estimating future land-use change. In that case, it might be necessary to split the model into sub-components to find the cause of the instability and thus validate its sub-components first.

3.4 Scale Issues in Land-Use Modelling

Scale and land-use modelling are interwoven phenomena. Choice of scale of modelling is of influence on the result as was described by Benenson (2007), who discussed the theoretical impact of scale on CA transition rules. Wu and Li (2006)

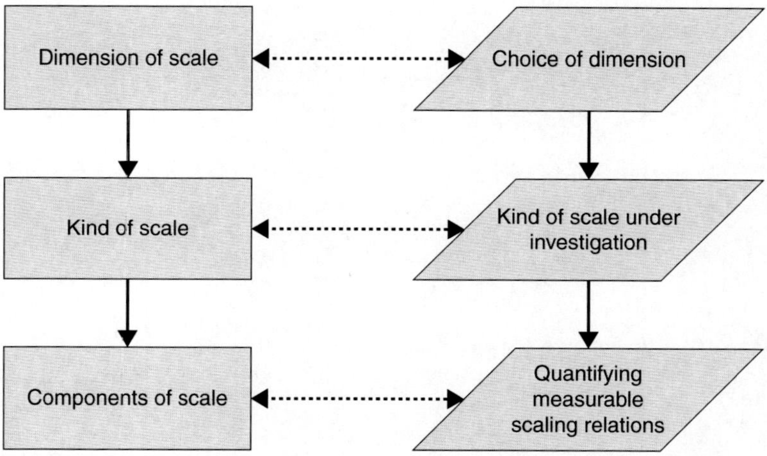

Fig. 3.5 Hierarchy of scale (after Wu & Li, 2006)

distinguish a conceptual hierarchy of scale (see Fig. 3.5) consisting of three levels: dimension, kind, and components of scale. Each of these levels is briefly described below.

The first level is the *dimension* of scale, which can be temporal, spatial or institutional (Bürgi, Hersperger & Schneeberger, 2005; Dungan et al., 2002; Wu & Li, 2006). Temporal scale is measured in time steps (e.g. years), spatial scale is measured in resolution (e.g. kilometres) and institutional scale is measured in institutions (e.g. two countries). Institutional scale is a special kind of scale since two institutions are never of the same size. Furthermore the institutional scale is discrete, whereas time and space are continuous.

On the second level of scale, Wu and Li (2006) distinguish different *kinds* of scale; see the left-hand side of Fig. 3.6. The intrinsic scale is the scale on which a process operates in reality. The process is observed by humans on an observation scale. When a certain process is modelled the modeller has to choose a modelling scale, which can be, but is not necessarily, the same as the observation scale. The modelling results are then presented to the policy-makers, who will look on a policy scale at the process. For explanatory purposes, the kinds of scale are related to land cover and land-use change in Fig. 3.6. This figure distinguishes between the land cover that can be observed (e.g. building) and the use to which the land is put (e.g. residential or commercial). Analysis of spatial developments normally starts with observation of land cover change processes using remote sensing (with a resolution that can range from 1 m up to 30 m). The analysis or modelling scale is normally an aggregate of the observation scale. In case of the Dutch CBS data used in the Land Use Scanner this is a resolution of 25 m × 25 m. The policy scale depends strongly on the focus of the policy theme and may involve a variety of scales. Our focus in this chapter is the process of land-use change, which is often more complex than the process of land-cover change because of the human factor. Different core principles (e.g. neighbourhood interaction) exist to describe the way land-use change processes

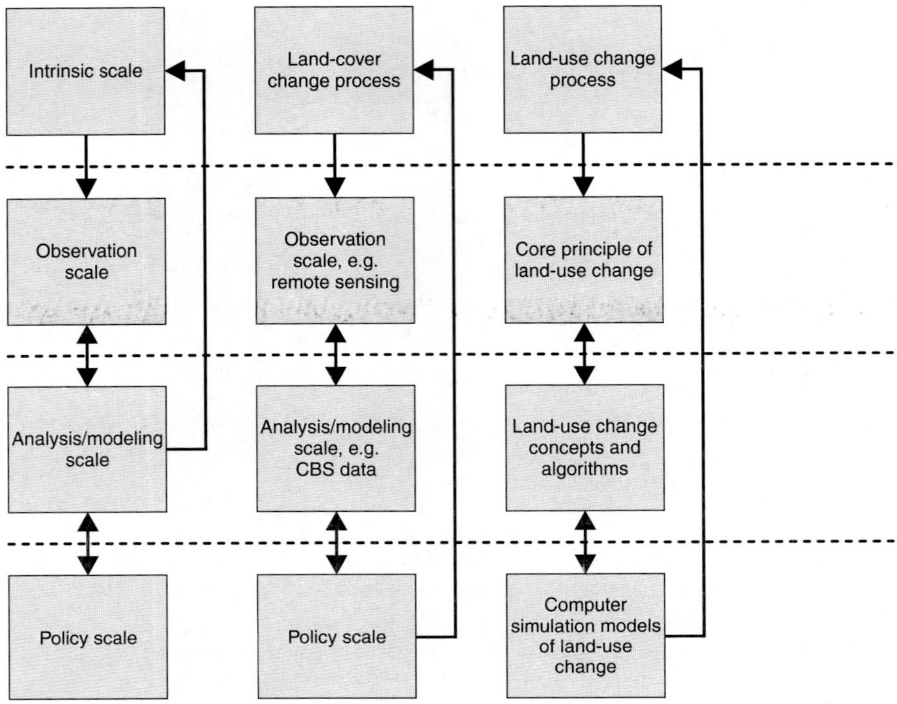

Fig. 3.6 Kinds of scale in modelling land-use change (after Wu & Li, 2006)

are supposed to have taken place. These can be seen as the observation scale of land-use change. Land-use change concepts and the algorithms that express them can then be thought of as the modelling scale of the land-use change process. The policy scale can be found in the outcomes of the models. Since policies influence processes of land-use change, there are feedback links back to the intrinsic scale, i.e. the land-use change process.

The third and last hierarchical level of scale relates to its *components* and is about quantifying and developing measurable scaling relations. These include cartographic scale, grain, extent, coverage and spacing (Wu & Li, 2006).

Figure 3.6 also places the framework in Fig. 3.1 in perspective. The implications of scale issues are important in the case of land-use modelling (Parker et al., 2001). Scaling difference between the policy scale and the intrinsic scale should be as small as possible to avoid model-induced errors. If, for example, the modelling scale level chosen is too large, the actual process of land-use change may be overlooked.

In recent research, the focus (as far as scale issues are concerned) has been on spatial correlation of land use and land-use change (Overmars, de Koning & Veldkamp, 2003; Verburg et al., 1999). Analysis at different temporal scales is more difficult due to incomplete or inconsistent time series of the required spatial data, (e.g. Li & Yeh, 2001). The institutional level of scale differs per country and case study.

3.5 Conclusions and Discussion

The literature analysis we present in this chapter is a first step towards acquiring a better understanding of the usability of land-use models. Knowing the core principles of a model can help a modeller use it in the right way and help the policy-maker to understand the basis of the model. It makes the 'black box' – which models sometimes are to policy-makers (and even modellers) – less black. To illustrate the degree of complexity of advanced models, it is estimated that someone who is new to such a model needs at least 2–3 months before he or she is fully capable of understanding and working with the model.

Of the many models that exist, this report mentions only a few. Some models are clearly too complex, which makes them difficult to implement at other spatial locations, such as UrbanSim of which the implementation at the Environmental Assessment Agency illustrated the difficulty to use this model in the context of a different location and data availability.

Care should be taken not to judge models by their spatial performance alone. Over-fitting of models is a well-known trap. There is a balance to be found in calibrating a model for a certain spatial location and the usability of this calibration for other spatial locations. Research aims might also influence this. Perhaps the aim is to build a model that can be used immediately at different spatial locations, or perhaps it is to build a model that has to be calibrated for each spatial location. The first option, a model that can be used at different spatial locations, seems the most preferred one.

It is better to use simpler models because it is easy to explain what happens and the results are easier to explain. It seems that the complexity land-use modellers are trying to capture is not something that can be modelled. There are simply too many factors to be taken into account. Agent-based models focus strongly on process resemblance, artificial neural networks focus strongly on spatial resemblance. Changing the focus of the model will result in a loss of its strong points, so a trade-off has to be made in which certain models are used to optimal effect.

More care should also be taken in listening to policy-makers. What do they actually need? This is currently the focus of the work of several PhD students in the Netherlands.

It is our opinion that – at least for the Netherlands – using a model such as Land Use Scanner, which works with suitability, demand and supply, is a good approach, but not for creating beautiful maps. Rather, it should be used more with a technique such as variant–invariant region, as Brown, Page, Riolo, Zellner, and Rand (2005) show in their paper on path dependence in agent-based models of land use. This works as follows. Different scenario's can be analyzed in the computer simulation model. Maybe from the results it can be seen that certain developments will always take place (invariant regions) or that certain areas are high in demand and their future use depends on the chosen scenario (variant regions). Then bottlenecks for spatial planning are recognised and policy-makers can make their decision instead of the model making the decision for them.

3.6 Themes for Future Research

Inspired by the literature review, we have selected several themes that warrant future research:

- thematic classification (number of land-use classes);
- extent (size of the study area);
- resolution (cell size); and
- time step (discrete interval in time).

The relation between these themes is shown in Fig. 3.7. Extent and cell size are assumed to be strongly correlated; for a study area with a specific extent, a certain resolution will be optimal. If a study area, for example, only measures 4 km × 4 km, it does not make sense to use a cell size of 1 km × 1 km. Sizes of 100 m × 100 m or 25 m × 25 m would then be more appropriate to discern spatial patterns within the area. The thematic classification of land use is often related to the extent and resolution of the case study. With increasing diversity (increasing number of classes), the minimal extent for a specific cell size increases.

Time step is related to all three (extent of study area, cell size and thematic classification), but the chosen time step mostly depends on the choice of classification and the development or amount of land-use change in a study area. If there is little change over time, the ideal time step for modelling can be larger than when there is a lot of change. Intuitively, if there are small time steps there is room to simulate more diversity (more different classes) than if there are large time steps. Of course, data is scarce, so research is often limited to a time step dictated by data supply.

For each of the four themes described above we list a few research options. These options focus on the Netherlands as they are meant to help the further development of the Dutch LUMOS models.

Thematic Classification

The definition of a thematic classification is a basic step in developing a land-use simulation model. Does a model simulate two (e.g. built or non-built) or more land uses? And what is the maximum number of land-use classes that can be simulated given a certain cell size and extent of a study area? In other words, what would be a

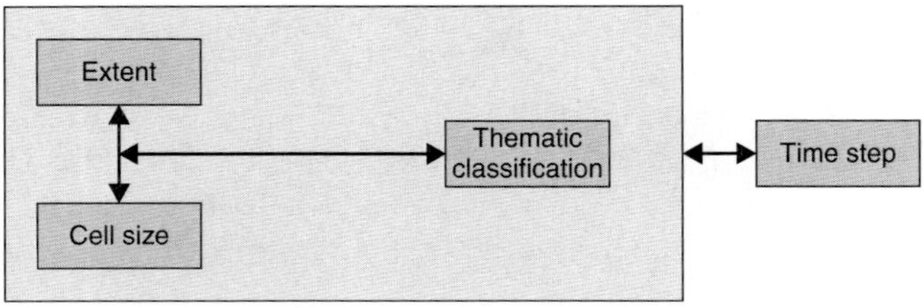

Fig. 3.7 Relation between cell size, extent of study area, thematic classification and time step

sound choice of the number of classifications to be used in land-use modelling given a certain area and cell size?

To start finding an answer the following option can be considered. First assume a certain extent of study area and cell size. Then the number of land-use classes is varied (from 2 up to 10). This can be investigated for different spatial locations. Although it may also be possible to approach this problem theoretically, i.e. with complicated mathematical equations, the ultimate goal is only to provide a rule of thumb with which modellers can determine an optimum number of classes.

Extent of Study Area

Often the choice of extent of a study area is arbitrary and data dependent. The bigger the study area (assuming cell size remains the same), the more land-use cells have to be allocated. Intuitively, there is a minimum extent that a study area should have before the use of a land-use model starts to make sense. A Markov model can be used to start off with a study area extent of four cells and only two classes (most extreme) and then, by making the extent bigger for each time step, an optimal extent can be found after which, statistically, the Markov model performs worse than not modelling (persistence).

Resolution (Cell Size)

For the Netherlands, further research could be done on the effect of using varying cell sizes (25/100/250/500/1,000 m) for different spatial locations in the country. The Land Use Scanner model offers the opportunity to simulate at all these resolutions, but the impact of these different choices has not yet been analysed systematically.

Time Step

To better understand ongoing spatial developments an analysis of historic developments is extremely useful. For the Netherlands several data sets are available for the years 1960–2000, with 10-year time steps. From these land-use maps, Markov matrices can be constructed. It can be investigated whether the stationarity principle holds or that the rate of change for certain classes changes over time. The matrix can be used to extrapolate land use into the future, validate the current model and obtain information on the impact of spatial planning regulations in the observed period.

References

Agarwal, C., Green, G. M., Grove, J. M., Evans, T. P., & Schweik., C. M. (2001). *A review and assessment of land-use change models: Dynamics of space, time, and human choice*. Bloomington, IN: South-Burlington, Center for the Study of Institutions Population, and Environmental Change, Indiana University.

Alcamo, J., Kreileman, G. J. J., Krol, M. S., & Zuidema, G. (1994a). Modeling the global society-biosphere-climate system: Part 1: Model description and testing. *Water, Air, & Soil Pollution, 76*(1–2), 1–35.

Alcamo, J. G. J., van den Born, A. F., Bouwman, B. J., de Haan, K., Klein Goldewijk, O., Klepper, J., et al. (1994b). Modeling the global society-biosphere-climate system: Part 2: Computed scenarios. *Water, Air, & Soil Pollution, 76*(1–2), 37–78.

Alonso, W. F. (1964). *Location and land use*. New Haven, CT: Harvard University Press.

Arentze, T. A., & Timmermans, H. J. P. (2000). *Albatross, a learning based transportation oriented simulation system*. Eindhoven: Eindhoven University.

Baker, W. L. (1989). A review of models of landscape change. *Landscape Ecology, 2*(2), 111–133.

Balmann, A. (1996). Farm-based modelling of regional structural change: A cellular automata approach. *European Review of Agricultural Economics, 24*, 85–108.

Batty, M., & Xie, Y. (1994). From cells to cities. *Environment & Planning B: Planning & Design, 21*, S31.

Bella, I. E. (1971). A new competition model for individual trees. *Forest Science, 17*(3), 364–372.

Benenson, I. (2007). Warning! The scale of land-use CA is changing! *Computers, Environment and Urban Systems, 31*(2), 107–113.

Borsboom-van Beurden, J. A. M., Boersma, W. T., Bouwman, A. A., Crommentuijn, L. E. M., Dekkers, J. E. C., & Koomen., E. (2005). Spatial impressions – Visualisation of future land use in the Netherlands. Bilthoven – The Netherlands, Netherlands Environmental Assesment Agency–550016003/2005.

Briassoulis, H. (2000). *Analysis of Land use change: Theoretical and modeling approaches*. Morgantown, WV: West Virginia University.

Brown, D. G., Page, S., Riolo, R., Zellner, M., & Rand, W. (2005). Path dependence and the validation of agent-based spatial models of land use. *International Journal of Geographical Information Science, 19*(2), 153–174.

Bürgi, M., Hersperger, A. M., & Schneeberger, N. (2005). Driving forces of landscape change – Current and new directions. *Landscape Ecology, 19*(8), 857–868.

Burnham, B. O. (1973). Markov intertemporal land use simulation model. *Southern Journal of Agricultural Economics, 5*(1), 253–258.

Chomitz, K. M., & Gray, D. A. (1996). Roads, land use, and deforestation: A spatial model applied to Belize. *World Bank Economic Review, 10*(3), 487–512.

Couclelis, H. (1985). Cellular worlds: A framework for modeling micro – macro dynamics. *Environment and Planning A, 17*(5), 585–596.

deMaris, A. (1992). Logit modeling: Practical applications, University of Iowa–07–086.

de Nijs, T. C. M., de Niet, R., & Crommentuijn, L. (2004). Constructing land-use maps of the Netherlands in 2030. *Journal of Environmental Management, 72*(1–2), 35–42.

Dungan, J. L., Perry, J. N., Dale, M. R. T., Legendre, P., Citron-Pousty, S., Fortin, M. J., et al. (2002). A balanced view of scale in spatial statistical analysis. *Ecography, 25*(5), 626–640.

Engelen, G., White, R., Uljee, I., & Drazan, P. (1995). Using cellular automata for integrated modelling of socio-environmental systems. *Environmental Monitoring and Assessment, 34*(2), 203–214.

Fearnside, P. M. (1996). Amazonian deforestation and global warming: Carbon stocks in vegetation replacing Brazil's Amazon forest. *Forest Ecology and Management, 80*(1–3), 21–34.

Ferrand, N. (1996). *Modelling and supporting multi-actor planning using multi-agents systems*. Santa Barbara, CA: Third NCGIA Conference on GIS and Environmental Modelling.

Fujita, M., Krugman, P., & Venables, A. J. (1999). *The spatial economy*. London: The MIT press.

Gardner, M. (1970). Mathematical Games: The fantastic combinations of John Conway's new solitaire game 'life'. *Scientific American, 223*, 120–123.

Geertman, S., Hagoort, M., & Ottens, H. (2007). Spatial-temporal specific neighbourhood rules for cellular automata land-use modelling. *International Journal of Geographical Information Science, 21*(5), 547–568.

Grimm, V., Berger, U., Bastiansen, F., Eliassen, S., Ginot, V., Giske, J., et al. (2006). A standard protocol for describing individual-based and agent-based models. *Ecological Modelling, 198*(1–2), 115–126.

Hagoort, M. (2006). The Neighbourhood Rules. Land-use interaction, urban dynamics and cellular automata modeling (300p). PhD, Faculty of geosciences, Utrecht University, Utrecht.

Heistermann, M., Muller, C., & Ronneberger, K. (2006). Land in sight?: Achievements, deficits and potentials of continental to global scale land-use modeling. *Agriculture, Ecosystems & Environment, 114*(2–4), 141–158.

Hilferink, M., & Rietveld, P. (1999). LAND USE SCANNER: An integrated GIS based model for long term projections of land use in urban and rural areas. *Journal of Geographical Systems, 1*(2), 155–177.

Hunt, J. D., Kriger, D. S., & Miller, E. J. (2004). Current operational urban land-use-transport modelling frameworks: A REVIEW. *Transport Reviews, 25*(3), 329–376.

Koomen, E., Loonen, W., & Hilferink, M. (2008). Climate-change adaptations in land-use planning; A scenario-based approach. In L. Bernard, A. Friis-Christensen, & H. Pundt (Eds.), *The European information society; Taking geoinformation science one step further* (pp. 261–282). Berlin: Springer.

Koomen, E., & Stillwell, J. (2007). Modelling land-use change; Theories and methods. Chapter 1. In E. Koomen, J. Stillwell, A. Bakema, & H. J. Scholten (Eds.), *Modelling land-use change; Progress and applications* (pp. 1–21). Dordrecht: Springer.

Koomen, E., J. Stillwell, A. Bakema, & H. J. Scholten, (Eds.). (2007). *Modelling land-use change. Progress and applications*. Dordrecht, The Netherlands: Springer.

Krugman, P. (1991). *Geography and trade (Gaston Eyskens lecture series)*. Leuven: Leuven University Press.

Krugman, P. (1999). The Role of Geography in Development. *International Regional Science Review, 22*(2), 142–161.

Kuijpers-Linde, M., Geurs, K. T., Knoop, J. M., Kuiper, R., Lagas, P., Ligtvoet, W., et al. (2007). Nederland Later, Tweede Duurzaamheidsverkenning, deel fysieke leefomgeving Nederland. Bilthoven.

Lambin, E. F., Rounsevell, M. D. A., & Geist, H. J. (2000). Are agricultural land-use models able to predict changes in land-use intensity? *Agriculture, Ecosystems & Environment, 82*(1–3), 321–331.

Lambin, E. F., Turner, B. L., Geist, H. J., Agbola, S. B., Angelsen, A., Bruce, J. W., et al. (2001). The causes of land-use and land-cover change: Moving beyond the myths. *Global Environmental Change, 11*(4), 261–269.

Langdon, W. B. (1998). *Genetic programming and data structures* (350p). MSc, University College, London .

Le Page, C., Bousquet, F., Bakam, I., Bah, A., & Baron, C. (2000). *CORMAS: A multiagent simulation toolkit to model natural and social dynamics at multiple scales*. Wageningen: Workshop 'The ecology of scales'.

Lesschen, J. P., Verburg, P. H., & Staal, S. J. (2005). Statistical methods for analysing the spatial dimension of changes in land use and farming systems, International Livestock Research Institute LUCC Focus 3 Office.

Li, X., & Yeh, A. G.-O. (2001). Calibration of cellular automata by using neural networks for the simulation of complex urban systems. *Environment and Planning A, 33*(8), 1445–1462.

Liao, T. F. (1994). Interpreting probability models. Logit, Probit, and Other Generalized Linear Models, University of Iowa–07–101.

Ligtenberg, A., Bregt, A. K., & Lammeren, Rv. (2001). Multi-actor-based land use modelling: Spatial planning using agents. *Elsevier, 56*, 21–33.

Lopez, E., Bocco, G., Mendoza, M., & Duhau, E. (2001). Predicting land-cover and land-use change in the urban fringe: A case in Morelia city, Mexico. *Landscape and Urban Planning, 55*, 271–285.

LUMOS (2005). Platform for land use modeling in the Netherlands. Lumospro. (2007). Project team website. www.lumospro.nl

Mas, J. F., Puig, H., Palacio, J. L., & Sosa-Lopez, A. (2004). Modelling deforestation using GIS and artificial neural networks. *Environmental Modelling & Software, 19*(5), 461–471.

Matthews, R., Gilbert, N., Roach, A., Polhill, G., & Gotts, N. (2007). Agent-based land-use models: A review of applications. *Landscape Ecology, 22*, 1447–1459.

McGarigal, K., & Marks., B. J. (1995). FRAGSTATS: Spatial pattern analysis program for quantifying landscape structure. Portland, OR: US, Department of Agriculture, Forest Service, Pacific Northwest Research Station-PNW-GTR–351.

Meyer, W. B., & Turner, B. L. (1992). Human population growth and global land-use/cover change. *Annual Review of Ecology and Systematics, 23*(1), 39–61.

Moeckel, R., Schurmann, C., & Wegener, M. (2002). *Microsimulation of land use. 42nd European Congress of The Regional Science Association*. Dortmund: Institut fur Raumplanung, University of Dortmund.

Muller, M. R., & Middleton, J. (1994). A Markov model of land-use change dynamics in the Niagara Region, Ontario, Canada. *Landscape Ecology, 9*(2), 151–157.

Nelson, G. C., & Hellerstein, D. (1997). Do roads cause deforestation? Using satellite images in econometric analysis of land use. *American Journal of Agricultural Economics, 79*(1), 80–88.

Overmars, K. P., de Koning, G. H. J., & Veldkamp, A. (2003). Spatial autocorrelation in multi-scale land use models. *Ecological Modelling, 164*(2–3), 257–270.

Overmars, K. P., de Groot, T., & Huigen, M. G. A. (2007). Comparing inductive and deductive modeling of land use decisions: Principles, a model and an illustration from the Philippines. *Human Ecology, 35*, 439–452.

Parker, D. C., Berger, T., & Manson, S. M. (2001). *Agent-based models of land-use and land-cover change. Adaptive agents, intelligence and emergent human organization: Capturing complexity through agent-based modelling*. Irvine, CA: LUCC International Project Office.

Pijanowski, B. C., Brown, D. G., Shellito, B. A., & Manik, G. A. (2002). Using neural networks and GIS to forecast land use changes: A land transformation model. *Computers, Environment and Urban Systems, 26*(6), 553–575.

Pijanowski, B. C., Pithadia, S., Shellito, B. A., & Alexandridis, K. (2005). Calibrating a neural network-based urban change model for two metropolitan areas of the Upper Midwest of the United States. *International Journal of Geographical Information Science, 19*(2), 197–215.

Pinto, N. N., & Antunes, A. P. (2007). Cellular automata and urban studies: A literature survey. *Architecture, City and Environment, 4*, 471–486.

Pontius, R. G., Boersma, W., Castella, J.-C., Clarke, K., De Nijs, T., Dietzel, C., et al. (2008). Comparing the input, output, and validation maps for several models of land change. *Annals of Regional Science, 42*(1), 11–37.

Pontius, R. G., Cornell, J. D., & Hall, C. A. S. (2001). Modeling the spatial pattern of land-use change with GEOMOD2: Application and validation for Costa Rica. *Agriculture, Ecosystems & Environment, 85*(1–3), 191–203.

Ricardo, D. (1817). On the Principles of Political Economy and Taxation, Library of Economics and Liberty.

Scopus. (2008). The largest abstract and citation database of research literature and quality web sources.

Silva, E. A., & Clarke, K. C. (2002). Calibration of the SLEUTH urban growth model for Lisbon and Porto, Portugal. *Computers, Environment and Urban Systems, 26*(6), 525–552.

Sinclair, R. (1967). Von Thunen and Urban Sprawl. *Annals of the Association of American Geographers, 57*(1), 72–87.

Skapura, D. (1996). *Building neural networks*. New York: ACM Press.

Tobler, W. (1979). Cellular geography. In S. Gale & G. Olsson (Eds.), *Philosophy in geography* (pp. 379–386). Dordrecht: Reidel.

Turner, B. L., Skole, D., Sanderson, S., Fischer, G., Fresco, L., & Leemans., R. (1995). Land-Use and Land-Cover Change Science/Research Plan, International Human Dimensions Programme on Global Environmental Change-7.

Veldkamp, A., & Fresco, L. O. (1996). CLUE: A conceptual model to study the conversion of land use and its effects. *Ecological Modelling, 85*(2–3), 253–270.

Verburg, P. H., de Koning, G. H. J., Kok, K., Veldkamp, A., & Bouma, J. (1999). A spatial explicit allocation procedure for modelling the pattern of land use change based upon actual land use. *Ecological Modelling, 116*(1), 45–61.

Verburg, P. H., van Eck, J. R. R., de Nijs, T. C. M., Dijst, M. J., & Schot, P. (2004). Determinants of land-use change patterns in the Netherlands. *Environment and Planning B: Planning and Design, 31*(1), 125–150.

Verburg, P. H., Schot, P. P., Dijst, M. J., & Veldkamp, A. (2004). Land use change modelling: Current practice and research priorities. *GeoJournal, 61*, 309–324.

von Thünen, J. H. (1966). *Isolated state: An English edition of Der isolierte Staat*. New York: Pergamom Press.

Waddell, P. (2002). UrbanSim: Modeling urban development for land use, transportation and environmental planning.

Wagner, P., & Wegener, M. (2007). Urban land use, transport and environment models. *disP, 3*, 45–57.

Walker, R. (2004). Theorizing land-cover and land-use change: The case of tropical deforestation. *International Regional Science Review, 27*(3), 247–270.

Walsh, S. E., Soranno, P. A., & Rutledge, D. T. (2003). Lakes, Wetlands, and Streams as Predictors of land use/cover distribution. *Environmental Management, 31*(2), 198–214.

Wear, D. N., & Bolstad, P. (1998). Land-use changes in southern appalachian landscapes: Spatial analysis and forecast evaluation. *Ecosystems, 1*(6), 575–594.

White, R., & Engelen, G. (1994). Cellular dynamics and GIS: Modelling spatial complexity. *Geographical Systems, 1*, 237–253.

White, R., & Engelen, G. (2000). High-resolution integrated modelling of the spatial dynamics of urban and regional systems. *Computers, Environment and Urban Systems, 24*(5), 383–400.

Wrigley, N. (1976). *Introduction to the use of logit models in geography*. Norwich: University of East Anglia–10.

Wu, J., & Li, H. (2006). Concepts of scale and scaling. In J. Wu, K. B. Jones, H. Li, & O. L. Loucks (Eds.), *Scaling and uncertainty analysis in ecology: Methods and applications*. New York: Springer, p. 6.

Part II
Practice

Chapter 4
A Sustainable Outlook on the Future of The Netherlands

Rienk Kuiper, Marianne Kuijpers-Linde, and Arno Bouwman

4.1 Introduction

In 2005, the Dutch Upper House asked the Dutch Government to prepare a highly integrated, long-term investment strategy in spatial planning that should pay more attention to the effects of climate change and make provisions for the further development of the Randstad conurbation. In response to this request, in 2007 PBL carried out the Second Sustainability Outlook. This study comprised two parts, each looking at sustainability from a different perspective:

1. *The Netherlands in a Sustainable World*; a spatial perspective analysing the relationship between the Netherlands and rest of the world (Hanemaaijer et al., 2008);
2. *The Netherlands in the Future*; a time perspective analysing the relationship between the Netherlands of today and the Netherlands in the future (Kuijpers-Linde et al., 2010).

The study presented in this chapter deals with the second, temporal perspective and addresses the question of the sustainability of the physical living environment in the Netherlands. In the present system, political and administrative decisions on the various social issues are almost invariably taken from a sectoral, and thus partial, viewpoint. This leads to partial solutions and compartmentalisation. This study shows that the pursuit of a sustainable Netherlands requires a more far-reaching integration of current policies. The planning of housing and employment areas, nature, landscape, infrastructure and energy supply are clearly related. These often conflicting activities can be accommodated much more effectively – while delivering the maximum possible quality of life to future generations – if they are looked at together. To accommodate the current demand for land, while ensuring

R. Kuiper (✉)
PBL Netherlands Environmental Assessment Agency, PO Box 30314, 2500 GH The Hague, The Netherlands
e-mail: rienk.kuiper@pbl.nl

E. Koomen, J. Borsboom-van Beurden (eds.), *Land-Use Modelling in Planning Practice*, The GeoJournal Library 101, DOI 10.1007/978-94-007-1822-7_4,

that future generations inherit a high-quality living environment, a more coherent, long-term vision is needed. The Outlook study shows how optimising the spatial allocation of activities can maximise the sustainability of the Netherlands. After a short elaboration of the different stages in this study, the main outcomes and the role of tools such as Land Use Scanner are discussed.

4.2 Researching Sustainability

4.2.1 Defining Sustainability

In the First Sustainability Outlook (MNP, 2004) the concept of sustainability was not easy to make concrete. Ultimately, sustainability was defined as the distribution of a certain quality of life and the possibilities for maintaining that distribution in the future. This quality of life is, then, determined by the availability of the resources needed to achieve the goals set. It was acknowledged, however, that the range of possible underlying objectives to choose from makes sustainability a heavy, value-laden concept. In the Second Sustainability Outlook, initially the endeavour was not only to examine the physical (i.e. planet) side of the problem, but, in collaboration with other policy assessments, also the long-term viability of the economic situation and the stability of social relations. Subsequently, however, it became clear that the concept of sustainability could only be transformed satisfactorily into workable principles for the planet side, in other words, the physical living environment.

As already indicated in our introduction to this chapter, to accommodate the current demand for land, while ensuring that future generations inherit a high-quality living environment, a more coherent, integrated long-term vision is needed. For our study the sustainability of the physical environment was broken down into the following six main themes:

1. *Climate change*: flooding risks and water damage, water shortages and salt-water intrusion;
2. *Biodiversity* (diversity of plant and animal life): connectivity and the quality of the National Ecological Network, compliance with the agreement of the European Commission to conserve internationally important habitats and species (Natura 2000 network);
3. A*ccessibility*: accessibility of cities, congestion on roads, unequal distribution of environmental impacts among different population groups;
4. *Quality of the living environment*: shortage of housing in both quantitative and qualitative terms (particularly location: green space in urban areas, rural living) and affordability;
5. *International business establishment*: availability of easily accessible business parks, presence of prime office sites, international hubs (particularly Amsterdam Airport Schiphol), and attractive residential areas;

6. *Landscape quality*: landscape values that are characteristic for the cultural-historic identity of the Netherlands and that are important in relation to tourism and recreation.

4.2.2 Drawing Up Two Reference Scenarios

After formulating these themes, current social trends were tracked to see whether existing goals were being achieved and what policy objectives remained to be realised in the future. The resulting picture was called the *Baseline* scenario, taking into account only policies that had been adopted by the Dutch or European Parliaments: it is a policy-neutral scenario. From this neutral perspective, past trends and patterns of spatial development were translated into maps depicting future spatial structures. This analysis was carried out for the continental Netherlands, focusing on spatial pressure, probable spatial dynamics and the impact of both. From this policy-neutral perspective, the demand for land given moderate economic and demographic growth was allocated with the Land Use Scanner model according to past patterns of, and trends in, spatial development. Average demographic and economic growth was assumed until 2040: more precisely, an economic growth of 1.7% per year and a population growth from 16.6 million people in 2007 to 17.1 million by 2040. This *Baseline* scenario, representing average spatial pressure, is in line with the OECD baseline scenario and was based on the Transatlantic Market (A2) scenario in the study 'Welfare, Prosperity and Quality of the Living Environment' produced by Dutch assessment agencies in 2006 (CPB et al., 2006). Because it was felt that the impact of future economic and demographic growth might be underestimated, an additional *High Development Pressure* reference scenario was formulated. It assumed annual economic growth of 2.1% and population growth from 16.6 million people to more than 20 million by 2040.

4.2.3 Interpreting Spatial Focuses

Future spatial developments were adjusted to obtain optimal land-use patterns for resolving a single persistent policy-related problem that is corresponding to one of the six main themes. The new picture of the desired use of space per sector that resulted was referred to as a 'focus'. A focus thus reflects a (partial) answer to one of the single problems. A focus is a spatial interpretation of a particular line of policy. Each focus not only analyses the spatial consequences of policy but also its effects on land-use patterns and land management. Graphic depictions of these spatial focuses are included in Chapters 1 and 5.

For each of the six focuses, land use was simulated – provided by Land Use Scanner – in such a way that specific, adverse developments were prevented. For each focus not only the spatial consequences of particular policies were analysed, but also the effects of a specific land-use pattern and land management. For that reason, the calculation of land-use dynamics was completed with estimates of the associated investment expenditures and maintenance and management costs. These steps were repeated for each focus. Finally, the outcomes of the land-use simulations for all focuses were compared with the reference scenarios. In this way the potential – spatial – conflicts and synergies could be highlighted. These insights formed the starting point for investigating alternative strategic actions and policy options.

4.2.4 Comparing Focuses with the Reference Scenarios

After the simulation of changes in land use and transport, the sustainability effects of all scenarios and focuses were assessed, both quantitatively and qualitatively. The two reference scenarios reflected business-as-usual conditions, while the focuses are optimised for one of the six sustainability themes. Nine indicators were designated that cover the six sustainability themes identified at the beginning of the study. Subsequently, additional effect-models, for example ecological models, combined with expert knowledge, were used to assess the impact on sustainability of all land-use simulations and corresponding transport intensities. Eventually these assessments were translated into three possible categories: an improvement of sustainability for that particular theme; a reduction of sustainability for that particular theme; and a stable situation for that theme, meaning no improvement or reduction.

4.2.5 Optimising Land Use in a Combination Map

Finally, the positive elements of the various focuses were combined as favourably as possible, so that ultimately a combined map (known, literally, as the *Combination Map*) was created in which the different focuses, and their land-use functions, were optimised for each of the three domains *people, planet* and *profit*. This map gives a picture of the Netherlands that is more sustainable and 'future-proof' and provides starting points for drawing up specific strategic options. Other forms of integrated maps that provide an even better combination of objectives may be conceivable, but these have yet to be found. The *Combination Map* is not a 'blueprint', rather it displays the best conceivable options at the moment.

In the end, the maps and indicators of the *Combination Map* were compared to both the *Baseline* scenario and the *High Development Pressure* scenario to identify what needs to be changed in current policies to achieve better performance on the six main themes of sustainability that had been identified at the beginning of the study.

In the remainder of this chapter, the main outcomes of the *Baseline* and *High Development Pressure* scenarios and of the *Combination Map* are discussed in more detail.

4.3 Land-Use Dynamics and Their Impact

The maps of current and future land use according to the *Baseline* scenario for the year 2040 are included in Chapter 1. As already mentioned above, the maps of the *Baseline* and *High Development Pressure* scenarios reflect actual trends in land-use dynamics and are based on modest and high economic and demographic growth respectively. The associated demands for land were provided by specialised housing, employment and agro-economic models. As these two reference scenarios are based on the trend-based spatial policies their outcomes differ only in the amount of projected urbanisation.

The main conclusions based on the outcomes of the simulations with the Land Use Scanner for the reference scenarios are:

- Land use in the Netherlands will change radically in the coming decades. The built-up area will increase by 15–26% by 2040 (the range covers differences in population and economic growth). The current total size of the built-up area is therefore expected to grow by about 25%.
- The majority of new housing will be sited in flood-prone areas; potential risk of flood-related damage will by 2040 be 2 to 3 times higher than today. In total the new built-up areas will account for about 25–30% of the economic value of all property in 2040. In the reference scenarios, the primary flood defences will have reached the required statutory standards by 2020. These improvements in safety levels will reduce the risk of flood-related damage in 2020 by a factor of 1.7 and the casualty risk by a factor of 3, as compared with the year 2000.
- Given the assumptions made about social trends and the effectiveness of existing policy, new housing development will take place near the major cities of the Randstad and in the province of North Brabant, the result of high demand for housing in those regions and an urban compaction policy in spatial planning.
- In the decades ahead, demand for new commercial sites will be concentrated mainly on the Randstad. The northern wing of the Randstad, in particular, will become part of a global network of cities. Many new commercial sites will be developed in this part of the Netherlands because the growth of jobs will be highest there.
- Because under the reference scenarios it is assumed that existing provincial spatial plans will be realised and that accessibility via the main road system is a significant business location incentive, new employment locations will be created and existing ones expanded, particularly in the centres of the large cities and their environs. Haarlemmermeer, the Zuidplaspolder, and the area to the south of Groningen are examples of locations that will be subject to these processes.
- In the reference scenarios, total mobility will increase by approximately 15–30%, with car use rising by 35–45%, between 2002 and 2040. The ranges allow for differences in population and economic growth. Accessibility of employment locations by car will improve by about 10% during this period, owing to

Table 4.1 Overview of sustainability impacts of each focus compared to the Baseline Scenario. A '+' indicates an amelioration of the baseline level of sustainability, '–' a reduction and '=' a more-or-less stable situation. The symbols in the Baseline scenario column indicate whether the anticipated trend-based developments meet the current policy objectives (+) or not (–), '=' indicates limited change and 'n.o.' indicates that no explicit policy objectives are defined

Sustainability indicator	Explanation	Baseline scenario	Impact focus versus baseline scenario					
			Climate change	Biodiversity	Accessibility	Quality of the living environment	International business establishment	Landscape quality
Protection against flooding	Risk of damages and human casualties	+	+	=	=	=	=	=
Adaptation to climate change	Potential areas for inundation within risk areas	+	+	+	=	–	=	=
Biodiversity	In Natura 2000 areas	–	=	+	=	–	=	+
Accessibility	Social financial accessibility benefits	n.o.	+	=	+	–	+	=
Quality physical living environment	Greenery surrounding the city and noise pollution	=	=	+	–	=	+	+
Conditions international business establishment	Congestion, perception of risks, growth in Northern Randstad area	n.o	=	=	=	–	+	=
Spacious and green living	Low density housing, close to amenities and within attractive landscapes	–	–	–	–	+	+	–
Quality of the landscape	Main qualities, perception and recreational values of the landscape	–	+	+	+	–	–	+
Spatial segregation	Spatial distribution of the various income brackets	n.o.	=	=	=	=	+	=

investments in the road system and job creation. After 2020, however, the accessibility of employment sites – without additional investment and pricing policies – will decline, owing to increased congestion and a decrease in the number of jobs.

- The proportion of homes experiencing noise levels of more than 55 dB will have increased from 43% today to 46% by 2040.
- In the reference scenarios, the National Ecological Network will be realised by 2018 (approximately 730,000 ha). The nature conservation areas in the network will remain fragmented, however. Environmental conditions will not improve sufficiently, either. Biodiversity goals will, therefore, not be achieved.
- In the future, the open landscapes so characteristic of the Netherlands will largely disappear, because of ongoing urbanisation.

As already mentioned in the description of the design of the reference scenarios (Section 4.2), Land Use Scanner was used to create explicit maps of future land use for the focuses, each of which represents one of the six sustainability themes for the physical living environment. For a more detailed description of these land-use maps, see Kuijpers-Linde et al. (2010). Here, it suffices to say that all land-use maps, trend-based scenarios, and focuses were assessed for their impact on sustainability. Table 4.1 shows the outcomes of these analyses; comparing the sustainability assessments of the six focuses with the *Baseline* scenario.

4.4 Optimising for Sustainability

This section highlights the integration of each of the six sustainability themes (or focuses) in the *Combination Map*. Figure 4.1 gives an overview of all relevant policies per sustainability theme. Their elaboration in the *Combination Map* and the assumptions made on the use of geographical space in the land-use simulations are discussed below in more detail for each sustainability theme.

4.4.1 Climate Change

The *Climate change* focus shows that the Netherlands can probably withstand climate change and rising sea levels for centuries to come and that structural measures, such as locating capital investments on the higher ground or substantially widening the coastal zone, are therefore not urgently required. The *Combination Map,* thus, assumes further investment in the low-lying areas of the Netherlands, particularly in the Randstad. The Map confines itself to a targeted differentiation of safety levels to reduce the damage and risk of flood-related casualties while creating a robust protection system on the Rhine-Meuse floodplain. Areas with the lowest safety standards will be kept free of new urban development as much as possible. The introduction of overflow dykes is expected to increase the predictability of any flooding and reduce the risk of casualties even further.

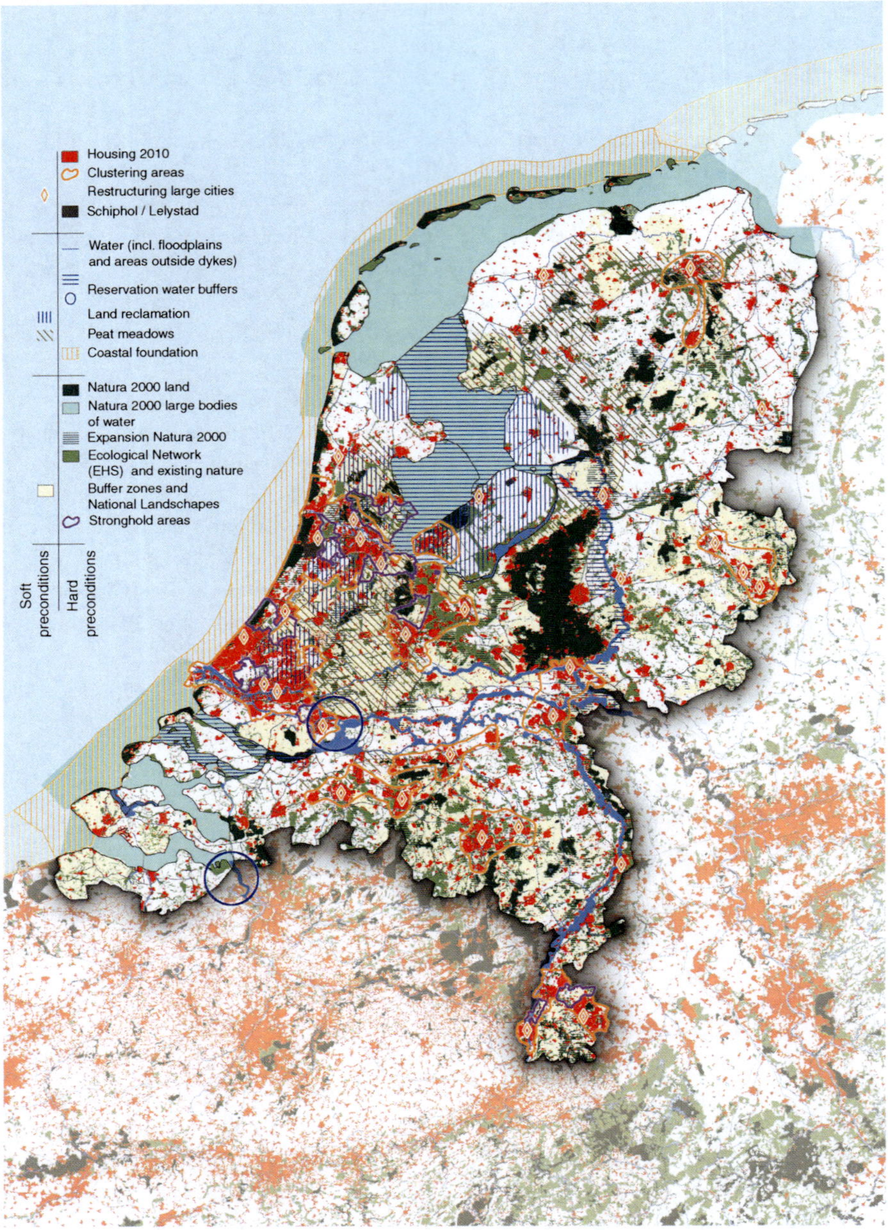

Fig. 4.1 Preconditions for the combination map

Although there are many uncertainties surrounding the rate and scale of climate change, and with it a rise in sea levels, the analysis presented in Chapter 2 shows that the diminishing opportunities for unhindered discharge of water from its major rivers determine the long-term future of the Netherlands. In the event of a two-metre rise in the sea level, other structural solutions may have to be found to accommodate main and peak discharges from the Rhine. However, two or three centuries would be needed for a rise in the sea level to occur that approaches the upper end of the KNMI's estimates. The densely populated lower reaches of the Rhine and Meuse rivers, with cities such as Rotterdam and Dordrecht on their fringes, are particularly vulnerable (see Fig. 5.1). To keep open long-term options for adjusting river discharges and storing water, the *Combination Map* contains areas reserved for water retention in the southwestern delta area, areas flanking the courses of the major rivers, and the IJssel valley and IJsselmeer areas. These emergency water-retention areas make the Netherlands more resilient in the event of any unexpected acceleration in sea level rise during this century.

Low-lying areas of the Netherlands must include areas for extra water storage. Some of the deepest polders (reclaimed land), or sections of these polders, as included in the *Landscape quality* focus, would be the most appropriate locations, because this would also serve to reduce the intrusion of salt water into the polder drainage system, and combat desiccation in surrounding nature conservation areas, as well as create additional opportunities for recreation and green residential areas. A separate saltwater drainage system would deliver additional options for expanding facilities for water-based recreation. Given the fact that options for adapting sewerage systems and reserving land for water retention at a later date are limited and would also entail higher costs, it has been assumed that new urban areas will have a robust design that includes provisions for additional water storage. This is also an important requirement for the restructuring of existing urban areas. Reservation of areas in the IJsselmeer area for longer term water-retention and the development in this area of internationally important nature conservation habitats would keep options open for substantial fluctuations in the water-table in that area and also reduce vulnerability to drought.

4.4.2 Biodiversity

The *Biodiversity* focus indicates that to comply with the Directive of the European Commission to conserve internationally important habitats and species (Natura 2000 network), the Netherlands needs to strengthen several nature conservation areas by enlarging them and reducing the environmental pressure exerted by the surrounding areas. The *Combination Map* incorporates this expansion of the Natura 2000 areas, as well as measures (including financial compensation) to be taken in the buffer zones around those areas. Greater emphasis is placed both on wetlands (peat marshes, regional river/stream systems, major water bodies and river areas) and the fringes of the Veluwe and the Utrechtse Heuvelrug.

4.4.3 Accessibility

The *Accessibility* focus shows that with a high degree of clustering and intensification, accompanied by the introduction of road-use pricing (based on time, place and environmental criteria), the investments needed to achieve the accessibility goals that have been set are lower than if only the infrastructure is expanded. Improvement in public transport also delivers a significant gain in accessibility in addition to that resulting from the introduction of road-use pricing. Both instruments have therefore been incorporated in the *Combination Map.*

Another advantage of clustering and intensification is that urban development takes up less space, which leaves the flexibility of the spatial structure of the Netherlands intact, makes it easer to reserve land for flood protection (ability to adapt to climate change) and limits impacts on the landscape. This makes the concentration of urban development a fundamental principle underlying the *Combination Map.* Clustering and intensification also have negative effects. The main ones are a (relative) decrease in the area of green space in and around the cities (recreational opportunities) and an increase in noise nuisance. As compensation, the Map takes into account the creation of additional areas of surface water and green space around the cities as part of the *Landscape quality* focus.

4.4.4 Quality of the Living Environment

The focus of *Quality of the living environment* performs less well than the *Baseline* scenario against many indicators, because of the greater claims on land and the development of urban land uses in attractive landscapes. These are, therefore, not included in the *Combination Map.* To improve the quality of the living environment, not only in new development areas, but also in existing housing areas, the *Combination Map* includes additional areas of water and green space around cities (see the focus of *Landscape quality*).

There is more government intervention reflected in the *Combination Map* than under the *Baseline* scenario, and urban development is prohibited in attractive residential landscapes, such as the National Landscapes and urban buffer zones. Consequently, less land is available for this type of housing development.

4.4.5 International Business Establishment

The focus of *International business establishment* assumes that only the northern wing of the Randstad is equally attractive to foreign businesses as the top-ranking cites of Europe (Barcelona and Munich). A great deal of the urban area in the western part of the Netherlands is, therefore, concentrated around Amsterdam. However, this has an adverse effect on other parts of the city and on the natural habitats, landscapes and water bodies around Amsterdam. This concentration around Amsterdam has not been incorporated into the *Combination Map* because there is unlikely to be such a bias towards Amsterdam in government planning. The focus of *International business establishment* also shows that moving part of the capacity of Amsterdam Airport Schiphol to Lelystad Airport (a slight shift to the northeast) would improve the quality of the living environment around Amsterdam and, on balance, the Netherlands as a whole. This relocation of airport capacity is, therefore, incorporated in the *Combination Map.*

4.4.6 Landscape Quality

The focus of *Landscape quality* assumes a restrictive urban development policy for National Landscapes and urban buffer zones this is included in the *Combination Map.* Additional green space around the cities and additional water bodies are included in the Map to reduce the negative effects of clustering and intensification strategies applied in the accessibility focus.

The extra emphasis on agricultural landscape management in the focus of *Landscape quality* is included in the *Combination Map.* This strategy is applied to National Landscapes, the peat meadows, a 5-kilometre zone around the larger cities and their urban buffer zones. The basic principle here is that the reform of the EU's Common Agricultural Policy will enable a substantial shift away from the present system of agricultural subsidies to a system of rewards for landscape stewardship by farmers in the public interest. This will make it possible to fund agricultural landscape management in those areas, as well as additional environmental measures in the buffer zones around Natura 2000 areas. The *Combination Map* also incorporates the greater degree of concentration of intensive forms of agriculture (greenhouse horticulture, intensive livestock farming) that are part of this focus.

4.5 Comparing Combination Map with Reference Scenarios

To identify where land-use developments should take place that differ from current trends, the land-use patterns and indicator scores of the *Combination Map* were compared to those of both the *Baseline* scenario and the *High Development Pressure* scenario. Figures 4.1 and 4.2 present the outcomes of this analysis with respect to urban growth, which is essential for many sustainability themes. Figure 4.2 shows the optimal spatial configuration of new urban built-up areas, taking into account the constraints and requirements of all the themes (focuses) integrated in the *Combination Map* but based on the demand for land as assumed in the *Baseline* scenario

Figure 4.3 shows the optimal spatial configuration of new urban built-up areas based on the demand for land under the *High Development Pressure* scenario that assumes higher economic growth rates and a population of almost 20 million people by 2040. When comparing the indicators of the *Baseline* scenario with those of the *Combination Map*, the *Combination Map* appears to perform much better than the *Baseline* scenario for most indicators.

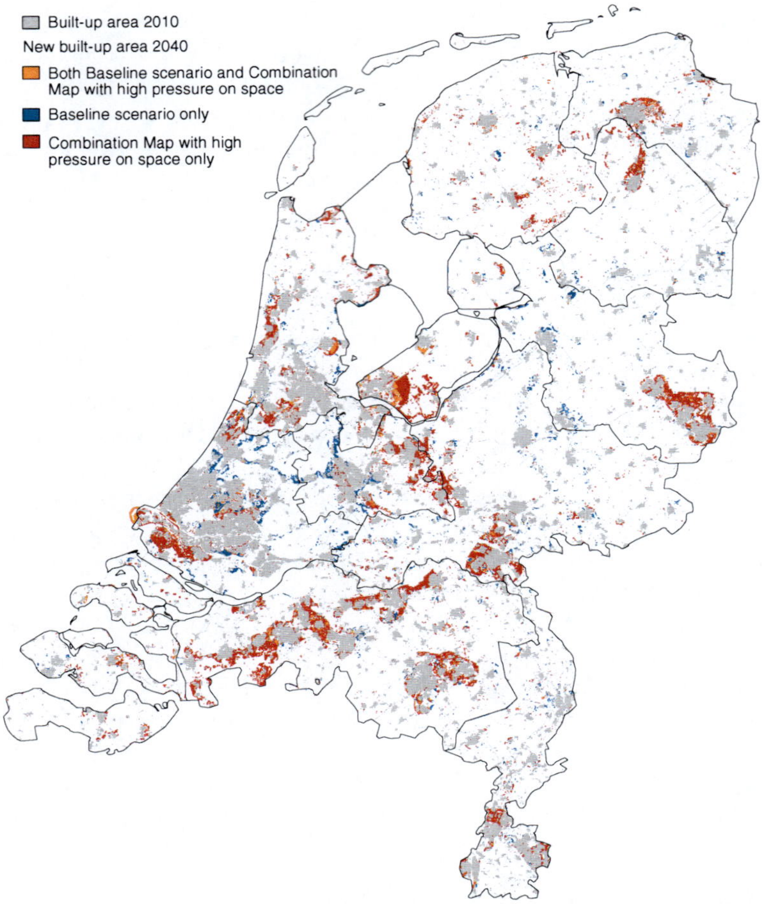

Fig. 4.2 Increase in built-up area according to the combination map compared to the Baseline scenario

The effects of the *Combination Map* score better on many indicators than the *Baseline* scenario does (see Table 4.2). However, not surprisingly, in the *High Development Pressure* scenario, the indicator scores would be significantly lower.

4.6 Conclusion

The land-use simulations and associated impact assessments performed for the Second Sustainability Outlook lead to the following main conclusions:

- To provide the necessary new housing, employment, transport infrastructure and green space and at the same time maintain the quality of the living environment for existing and future generations, a more holistic approach to development and

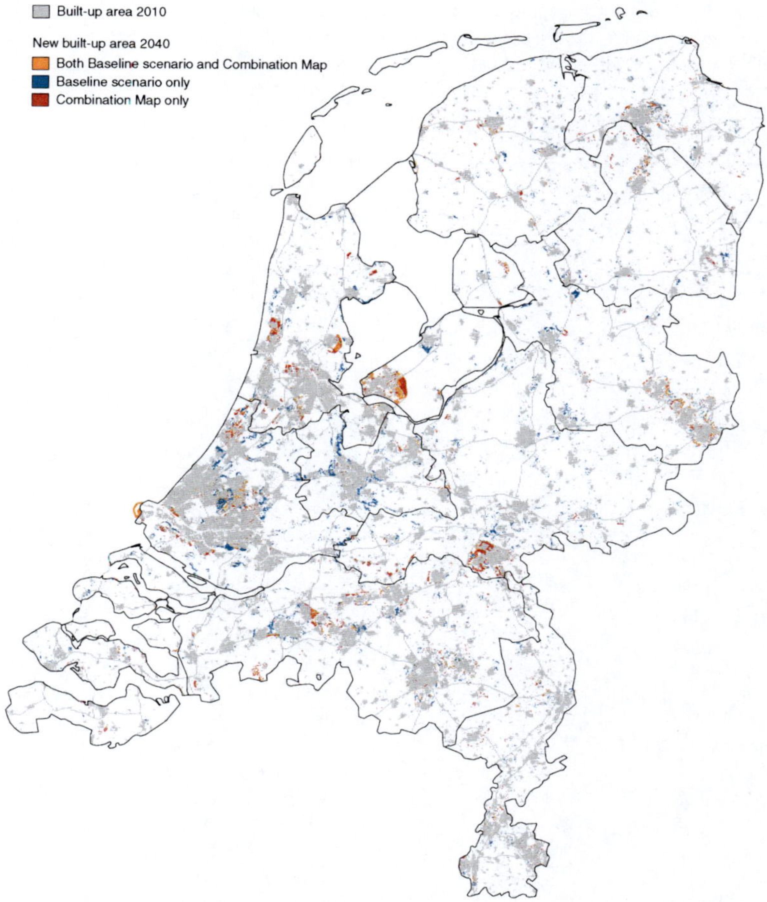

Fig. 4.3 Increase in built-up area according to the Combination Map compared to the High Development Pressure scenario

the environment is needed. Only then will it be possible to meet all the policy objectives and create a spatial structure that can rightly be called sustainable.

- The greatest improvements can be achieved by ensuring a much better fit between urban development and infrastructural measures and a better fit between flood protection, habitat creation/restoration and landscape development. Important gains can also be made through integrated approaches for agriculture and nature and landscape quality – and even integrating measures for flood protection with those for improving the attractiveness of the Netherlands to foreign businesses.
- Existing policy objectives provide sufficient opportunities for taking concrete steps towards achieving more sustainable use of land in the Netherlands. However, realising this will depend on the vigorous implementation of these policies at the local level and on effective harmonisation with EU policy.

Table 4.2 Effects of the Combination Map (development pressure on land in line with the Baseline scenario) compared with the Baseline scenario. The symbol '+' denotes a better performance of the combination map, while '=' indicates similar scores for both

Sustainability indicator	Combination map	Effects
Protection against flooding	+	Differentiation in safety levels, use of overflow dykes, concentration of urban development in low-lying areas of the Netherlands with the highest safety levels and limited urban expansion in areas flanking the major river
Adaptation to climate change	+	Designation of specific areas for future flood retention within flood-risk areas and more water bodies in and around the cities to provide more options for water management
Biodiversity	+	Expansion of Natura 2000 areas, additional green around cities, and landscape management of buffer zones National Landscapes
Accessibility	=	Improved accessibility resulting from urban compaction policy brings homes and employment closer to each other
Quality physical living environment	+	More green and water around cities and an improvement in the quality of agricultural landscapes; less noise nuisance from Amsterdam Airport. Urban intensification increases pressure on the quality of the living environment
Conditions international business establishment	+	More green and water around the cities, and improvements in the quality of agricultural landscapes; less noise nuisance from Amsterdam Airport
Spacious and green living	=	In and around the Randstad, there is some scope for developing new green living environments
Quality of the landscape	+	Areas with high landscape quality kept free of urban development and agricultural intensification; more water and green around the cities, more management of the agricultural landscape and reorganising of dispersed greenhouse horticulture and intensive livestock farming
Spatial segregation	=	Effect barely differs from structural trend

Clear spatial planning policies and enforcement of adopted land-use plans are important requirements for creating a sustainable living environment. The new Spatial Planning Act allows the preparation of spatial visions that can prioritise the spatial opportunities and conflicts in the case of developments of national importance, accompanied by an enforceable implementation agenda. From a sustainability perspective, benefits that will accrue in the more distant future should weigh more heavily when political decisions are taken.

- Clustering and intensification, as applied in the accessibility focus, will deliver considerable accessibility benefits. If the rise in mobility and congestion levels is modest, these gains will be greater than those that would be realised through the investments in the road network proposed by the Mobility Policy Document. Investment in infrastructure is more efficient if it is made according to a sequential approach: spatial policy (clustering and intensification) – road-use pricing – physical expansion of infrastructure.
- The Randstad will increase in greatly in size, particularly under the *High Development Pressure* scenario. The addition of new urban areas will then form a wider ring than is currently (2010) the case: running from Amsterdam to Leiden, The Hague and Rotterdam, via the Brabant linear conurbation to Nijmegen, Arnhem and on through Ede, Amersfoort and Utrecht to Almere and Amsterdam.
- In the *Combination Map*, some of the greenhouse horticulture complexes now situated in the western part of the Netherlands will move to the fringes of the Randstad (south of Rotterdam). To alleviate pressure on available space in the west, consideration could be given to designating areas elsewhere in the Netherlands for greenhouse horticulture.
- There are many opportunities for combining creation of habitat with spatial policies for the river areas (land reserved for widening the river IJssel and for the Kampen and Dordrecht bypasses), the IJsselmeer area and the eastern half of the Green Heart. Possible measures in the latter region include: inundation of the lowest-lying polders (reclaimed from lakes); limited flushing with fresh water to remove intruded salt water; and peat marsh development.
- Clear spatial planning policies that will bring the price of agricultural land under control, are a necessary precondition for ensuring the continuity and development of land-based agriculture that is needed to manage the cultural landscapes (National Landscapes) and for providing effective buffer zones around Natura 2000 areas. Financial compensation is necessary for agricultural landscape and environmental management in all these areas. The necessary conditions can be met through a combination of EU agricultural subsidies and national policies for the National Landscapes and Natura 2000 areas. The reform of the EU's Common Agricultural Policy (CAP) in 2013 is an essential, although uncertain factor here. Co-funding will be needed and will have to come from national as well as EU sources.

4.6.1 Policy Actions for the Short and Long Term

Both short-term and long-term policy initiatives can be used to achieve greater integration of sectoral policies, which in turn will deliver sustainability gains. Implementing these initiatives will not require any fundamentally new policies. The National Spatial Strategy and many other policy documents already contain many policies that lean in this direction. The National Spatial Strategy Monitor does show, however, that we cannot always be sure that policy objectives can be achieved in practice. Indeed, the Second Sustainability Outlook also demonstrates that stronger policies will be needed, particularly for flood safety in the longer term and to meet international nature conservation commitments.

4.6.2 Merits of Land Use Scanner

As has already been noted, cohesion at various levels and between various parts plays a central role in discussions about the way forward to achieve a more sustainable future for the Netherlands. The Land Use Scanner model made it possible to literally 'map' the future spatial dynamics of the Netherlands, thus enabling analysis of these dynamics and their interrelations at a national scale, as well as an assessment of their impact. A major advantage above exploration of the future based solely on expert judgment is the quantitative character of the Land Use Scanner model. Land Use Scanner can allocate the demand for land within regions, which makes regional differences in spatial pressure literally visible. The decision to compute not only the effects of a *Baseline* scenario, but also those of a *High Development Pressure* scenario was advantageous because it made clear that spatial patterns of land use, in particular that of urban growth, are very different from the Baseline scenario.

Nevertheless, the quality of the outcomes clearly depends upon the quality of the input in terms of data on the demand for land, GIS data, scientifically validated rules for allocation, and expert judgment. For that reason, a continuous comparison of model outcomes with findings in the scientific literature and expert knowledge is imperative, with the result that the process of computation is quite labour intensive.

Another point is the improvement in the understandability of grid cell maps. In communication with policy-makers and stakeholders, it appears that sketches conveying only the main outcomes work better than detailed maps of grid cells.

The Netherlands is only a small country, but it still needs to find space for housing, employment and transport, while simultaneously maintaining the quality of the living environment and green spaces. The best way to use the available space as effectively as possible is to view these functions and features as a cohesive whole, including the additional water-management problems resulting from climate change. This study shows how optimising the spatial allocation of various land-use functions can maximise the sustainability of the Netherlands in the future.

References

CPB, MNP, & RPB (2006). Welvaart en Leefomgeving. Een scenariostudie voor Nederland in 2040 (Prosperity, wellbeing and quality of the living environment. A scenario study for the Netherlands in 2040) CPB/MNP/RPB, Bilthoven/The Hague.

Hanemaaijer, A., de Ridder, W., Aalbers, T., et al. (Eds.). (2008). *The Netherlands in a sustainable world: Second sustainability outlook*. Bilthoven: PBL Netherlands Environmental Assessment Agency.

Kuijpers-Linde, M., Kuiper, R., Geurs, K., Knoop, J., Lagas, P., Ligtvoet, W., et al. (Eds.). (2010). *The Netherlands in the future, second sustainability outlook, the physical living environment in The Netherlands*. Bilthoven: PBL Netherlands Environmental Assessment Agency.
MNP (2004). Quality and the future. Sustainability outlook. Bilthoven: PBL Netherlands environmental assessment agency.

Chapter 5
Coupling a Detailed Land-Use Model and a Land-Use and Transport Interaction Model

Barry Zondag and Karst Geurs

5.1 Introduction

As already described in the preceding chapter, the study 'The Netherlands in the Future; The Second Sustainability Outlook' (MNP, 2007) constructed alternative future spatial strategies for the Netherlands and evaluated them using a diverse set of sustainability indicators, such as flooding safety, biodiversity, accessibility and landscape protection. As in previous studies, Land Use Scanner was used here as an instrument to allocate demand for land of different land-use types within a region. At a regional level, this demand had been calculated earlier by sector-specific models for the Baseline scenarios. The analytical framework was further extended with the Tigris XL model, a Land Use and Transport Interaction model (Significance & Bureau Louter, 2007). The specific objectives for using the Tigris XL model in addition to Land Use Scanner were:

- To simulate transport endogenously and analyse the impacts of joint alternative spatial and transport strategies on transport indicators (such as congestion) and accessibility indicators (in both geographical and monetary terms);
- To simulate and analyse the mutual influence of land use on transport and transport on land use;
- To calculate the impacts of alternative land-use and transport strategies on the housing and labour market at a regional level.

Land Use Scanner and Tigris XL were applied for the first time together within one analytical framework in 'The Netherlands in the Future' project. The interaction between the models was set up within the project by following a pragmatic approach that used the strength of both models without making fundamental changes to either of them. Generally speaking Land Use Scanner provides more detail about the

B. Zondag (✉)
PBL Netherlands Environmental Assessment Agency, PO Box 30314, 2500 GH The Hague, The Netherlands
e-mail: barry.zondag@pbl.nl

E. Koomen, J. Borsboom-van Beurden (eds.), *Land-Use Modelling in Planning Practice*, The GeoJournal Library 101, DOI 10.1007/978-94-007-1822-7_5,

supply side of land by distinguishing more land-use types and contains detailed characteristics related to location, for example spatial planning policies. Apart from the incorporation of transport, Tigris XL provides more detail to inform its demand modelling, for example by simulating choice of location by different household types and economic sectors.

However, to be able to successfully link both instruments within one analytical framework, it is important that the consistency between Land Use Scanner and Tigris XL increases in terms of classifications, spatial units, and time horizons used. To improve the consistency, input data for the two models was harmonised as far as possible. For some data categories, an adjustment of the classification is also needed. In addition to harmonizing the data, alternative ways of interaction between the two models were tested to see how the models might make best use each other's merits.

Following this short introduction, a short description of the Tigris XL model and its characteristics is given in Section 5.2. For a description of Land Use Scanner reference is made to Chapter 1 of this book and previous papers (e.g. Hilferink and Rietveld, 1999). Section 5.3 describes how Land Use Scanner and Tigris XL have been linked. A description of an actual application of the combined framework follows in Section 5.4. In Section 5.5 conclusions are presented regarding the interaction between the two models, the contribution of the combined framework to policy-making and recommendations for further improvement of the framework.

5.2 Short Description of Tigris XL

In the period 2002–2005, a consortium of Significance (part of RAND Europe at the time) and Bureau Louter developed the Tigris XL model in a number of sequential projects for the Transport Research Centre in the Netherlands (part of the Ministry of Public Works, Transport and Water Management). These projects consisted of a combination of model development, model application and model testing. Tigris XL is a so-called Land Use and Transport Interaction (LUTI) model, which uses the National Transport Model (LMS) of the Netherlands as its transport module (RAND Europe & Bureau Louter, 2006a; Zondag, 2007).

The key characteristics of the Tigris XL model are:

- It is a dynamic land-use model simulating time steps of 1 year. This enables it to simulate path dependency and to analyse how the system evolves over time. The key reason for this approach is that, due to many different time lags in the system (economic cycles, planning cycles, construction time, etc.), a general equilibrium does not exist in land use. Within its incremental structure the model uses partial equilibrium conditions, for example to match supply and demand within a year on the housing market.
- The module for choice of residential location has been statistically estimated, for six different household types, based on a large housing market survey (over 100,000 respondents). The module for choice of location of firms (represented as jobs in the model) has been estimated at municipality level for seven economic sectors on time-series data for the period 1986–2000. This enables the model to:

  - estimate the relationship between location of jobs and location of residents. Tigris XL does not use a pre-assumed hierarchical relationship but rather one that, based on estimation results, varies between economic sectors and household segments;
  - statistically estimate, based of revealed preference data, the influence of accessibility on the distribution of households and jobs.

- The land-use and transport system is simulated in Tigris XL in a set of linked modules addressing specific aspects of the systems (e.g. demography). The overall architecture determines the sequences of the modules and data exchanges between the modules. This allows flexibility to change or re-estimate specific modules without the need to change the whole framework.
- The land-use model operates in a way tailored to the National Transport model (LMS) of the Netherlands. The spatial detail of the land-use model is at the level of transport zones.
- The model has a three-layer structure, namely land, objects (such as dwellings) and activities (people, jobs).
- The land-market module in Tigris XL has different options to simulate the influence of government on spatial development. Depending on its settings it can simulate new urban development in three ways:

  - as a free-market development following the preferences of residents and firms. Location choices are only restricted by the lack of available land or possible spatial planning restrictions (such as nature areas);
  - as regulated development via designated locations and numbers of houses (in this case only the location choice of residents is simulated with the model); or
  - as an intermediate variant that takes development plans as a starting point but adjusts them within a certain range to actual market demand.

Tigris XL consists of five modules, which together address demography and spatial markets. Figure 5.1 presents an overview of the model and the main relationships between the modules. The model distinguishes two spatial-scale levels: the municipality level (approximately 450 regions) and local transport zones of the National Model System (LMS sub-zones; 1,308 sub-zones cover the Netherlands).

Core modules in the model are the housing market and labour market modules. These account for the effect of changes in transport on residential or firm location behaviour and in this way link changes in the transport system to changes in land use. A land and real-estate module simulates supply constraints arising from the amount of available land, land-use policies and construction. The module defines different levels of government influence on spatial development, ranging from completely regulated towards free market, and various feedback loops between demand and supply are also available. A demographic module is included to simulate demographic changes at the local level. At the national level, the model's output is consistent with existing demographic and socio-economic forecasts for population, labour force, income levels and employment.

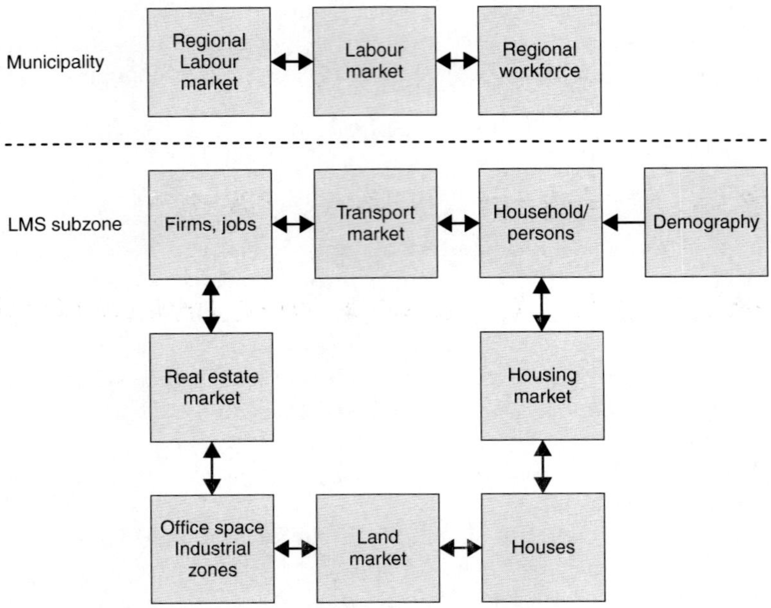

Fig. 5.1 Functional design of the Tigris XL model

The demographic module addresses processes such as births, deaths and ageing of the population, as well as changes in the composition of households. It deals with persons by category (gender, age) as well as households by category (size, income, etc.). The demographic module operates at the local subzone level and also processes the spatial distribution of scenario-based international migration flows.

The *land-market and real-estate market module* processes changes in land use and buildings, office space and houses, and addresses both densification of houses or floor space within existing urban areas and green-field developments. The changes within the existing urban area are exogenously specified and only administrated within this module. The simulation of the land-use changes in non-urban land can be modelled exogenously or endogenously depending on the setting for the level of market regulation. As said before, this can vary from a regulated land-use planning system, using exogenous input on the size and location of development sites, to a non-regulated market endogenously calculating the size and location of development sites.

The aim of the *housing market module* is to simulate annual moving (if any) of households. The housing market module simulates two choices: the choice to move or stay, and the choice of residential location following a move. The choice of residential location has a nested logit structure and incorporates a regional and a local scale level: there is a choice of region and a choice of location within the region. Choices depend on household characteristics, local amenities, prices, accessibility and distance (travel time) between the new and old residence. The parameters of the move/stay and residential location choice function, for each household type, have been estimated from a large four-annual housing market survey of more than 100,000 households in the Netherlands.

The *labour market module* in Tigris XL models changes in number of jobs by seven economic sectors and changes in workforce at a regional and zone level. For each sector, the influence of accessibility on the spatial distribution of employment has been modelled in combination with a set of other explanatory variables. The parameters have been estimated on a historical data set (1986–2000) including employment figures by sector at a municipality level. The labour market module interacts with the demographic, land and real-estate, housing market and transport modules.

The *transport module* calculates changes in transport demand and accessibility and is integrated with the National Transport Model (LMS). The LMS consists of a set of discrete-choice models for various choices in transport such as round trip frequency, mode of travel and destination, departure time and route. The LMS is based on micro-economic utility theory, enabling the derivation of utility-based accessibility measures.

Since its finalisation, Tigris XL has been used to evaluate a variety of transport and spatial policies. Examples of the application of the model for evaluating policy issues include the evaluation of a new road or rail infrastructure, transport pricing policies, land-use plans and urbanisation strategies (RAND Europe & Bureau Louter, 2006b; Significance & Bureau Louter, 2007, 2009; Significance et al., 2007) The model has been applied for the evaluation of the following issues:

- Effects of long-term socio-economic scenarios on transport and land use, including the translation of socio-economic trends and land-use plans into socio-economic input data for the transport model;
- Land-use effects of transport policies, including road and public transport infrastructure and operations, as well as pricing policies;
- Effects of alternative land-use policies on land use and transport:
  - for new residential or commercial development sites, the model can use different assumptions for the regulation of the land market;
  - urban densification strategies;
- Effects on land-use and transport of integrated land-use and transport strategies.

5.3 Linking Tigris XL and Land Use Scanner

This section describes how the coupling between Tigris XL and Land Use Scanner has been set-up. It discusses which information is exchanged, the conversion and/or processing steps and what modifications have been made to the Tigris XL model. First a brief overview is presented of the main differences and overlaps between the two models. This overview was needed to design the proper level of interaction between the models.

5.3.1 Overview of Differences and Overlaps

When Tigris XL and Land Use Scanner are compared it can be concluded that:

- Both models have different allocation mechanisms. In Land Use Scanner, regional claims for housing, labour, etc., are allocated to grid cells by a constrained logit model specifying the highest utility. The allocation is constrained by land-use claims provided at the regional level. The Tigris XL model allocates national projections to the regional and zone level by simulating the demand preferences of residents and firms. In both models, supply-side restrictions, depending on land characteristics and spatial planning, influence the location choices. This occurs in Tigris XL at the zone level and in Land Use Scanner at the grid cell level.
- Land Use Scanner uses exogenous land claims at a regional level as input and land use by grid cell (100 m × 100 m) as its output. This means that the model simulates only changes in land use within regions and not processes leading to land-use changes in other regions (e.g overflow effects from large cities to adjacent regions). Tigris XL uses national demographic and economic scenario projections as input and simulates both the inter-regional and intra-regional changes in population, employment and land use.
- The Tigris XL model simulates land-use changes resulting from urban development or government plans but does not simulate changes between different types of non-urban land-use functions. Land Use Scanner simulates changes in urban and several non-urban land-use functions, including agriculture and nature. The model also simulates non-urban transitions in land use.
- A nation-wide classification system exists to classify residential areas in the Netherlands by type of neighbourhood, for example, countryside or urban centre. In Tigris XL, the classification of residential area types is a fixed exogenous input per zone. Land Use Scanner calculates endogenously changes in residential area types for the cells (e.g. from non-central into central urban area). This means that in future years Land Use Scanner will classify the grids (and indirectly the zones they are located in) differently from Tigris XL (which uses percentage of residential area types at zone level).

The key characteristics of the two models are summarised in Table 5.1.

5.3.2 Set-Up of Interaction

For the 'Second Sustainability Outlook on the future of the Netherlands' project (MNP, 2007) Tigris XL and Land Use Scanner were loosely coupled (see Fig. 5.2). In this set-up the models operate individually and interactions between the models take place by exchanging input and output through data files. By setting up these interactions, some modifications have been made to Tigris XL to take advantage of data available in Land Use Scanner and to improve the consistency in data

Table 5.1 Key characteristics of land use scanner and Tigris XL at various spatial and temporal resolutions

Spatial and temporal resolution	Land use scanner	Tigris XL
National level	Uses scenario input directly in definition of local suitability and indirectly as input for the sector models calculating the regional claims	Uses scenario input
Regional level (e.g. 12 provinces, 40 COROP regions or over 400 municipalities)	Exogenous input of land claims for employment, housing and other sectors can be provided at any regional level	Provides an initial spatial distribution of residents and employment at level of municipalities
Zonal level (1,308 LMS sub-zones for the Netherlands)	Not applicable	Simulates spatial distribution of residents and employment, land-use change and changes in accessibility
Local level (100 m × 100 m grid cells)	Simulates land-use change for the cells using the regional claims and local (multi-criteria) suitability definition	Not applicable
Transport network	Static accessibility indicators are used to describe local suitability, no dynamic network is included	Calculates transport indicators such as travel times, number of vehicles and congestion
Temporal resolution	2010, 2020, 2040	Yearly until 2040

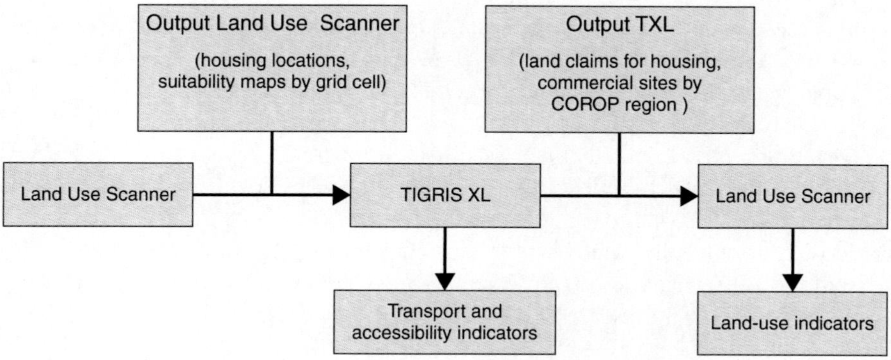

Fig. 5.2 Interaction between Land Use Scanner and Tigris XL (TXL)

classes between the models. In addition, post-processing procedures have been developed to transform Tigris XL output such as residents, houses and employment by sector into regional land-use claims, which are used as input data for Land Use Scanner.

The manner in which Land Use Scanner and Tigris XL are coupled depends on which type of housing market is assumed in the particular application. The housing market in the simulation can either be a regulated market, with fully planned development of all new housing construction, versus a more free market, consisting, for example, of a zoning policy and market-driven construction of houses. In the first case, the location or zone of green-field housing construction on agricultural land is determined exogenously in Tigris XL. Output of Land Use Scanner for 2020 and 2040 related to new residential areas is used as input data (in terms of hectares and number of houses) for Tigris XL. Land Use Scanner output data is processed by aggregating the cell data to the zone level and by distributing the simulated developments into time steps of 1 year. In the second case, the demand preferences of households, as simulated in the module for choice of residential location in the Tigris XL model, influence the location of housing construction. The resulting population developments also influence the location of new economic activities, which is calculated in the labour market module of Tigris XL. Both the residential and employment developments are then converted into land claims at the regional level and used as input for Land Use Scanner. Both types of coupling are discussed in more detail in Section 5.3.4.

5.3.3 Use of Land-Use Data and Restriction Maps

Several changes were made to the land-use classification and data of Tigris XL to improve consistency with Land Use Scanner and to take advantage of the detailed land-use information in the latter. The following changes were made:

- The land-use category of housing in Tigris XL was split into three residential area types similar to those used in Land Use Scanner. At the housing stock level in Tigris XL, a classification of five residential area types is used, which can be simply aggregated into the three categories of Land Use Scanner.
- The land-use type of agriculture in Tigris XL was split into agriculture and horticulture, based on data from Land Use Scanner. The two categories needed to be distinguished to reflect differences in the cost of acquiring land for urban development.
- Maps restricting the options for development in Land Use Scanner, for example those related to external safety, noise pollution, nature protection and hydrological constraints, were used in Tigris XL to exclude locations as options for development.

5.3.4 Interactions for Differing Housing Market Circumstances

This subsection describes in more detail two alternatives ways in which Land Use Scanner and Tigris XL have been linked. The preferred way of coupling the two systems depends upon assumptions made about housing market conditions. If the housing market is more regulated and thus supply-driven then the detailed supply information from Land Use Scanner plays a more dominant role, but if the housing market is more demand-driven then the demand modelling in Tigris XL plays a more dominant role. It is assumed that the location choices of firms will not be strongly regulated and in both of these Land Use Scanner–Tigris XL coupling options, the labour market module in Tigris XL has been used to calculate the spatial distribution of employment.

In cases of *regulated or planned housing supply*, the supply information on new residential locations as simulated per grid cell in Land Use Scanner for 2020 and 2040 are exported to fill the planning files of Tigris XL by year, zone and residential area type. The housing market module in Tigris XL uses houses as supply unit instead of hectares of residential land and the number of newly constructed houses by zone are calculated based upon the hectares of new residential construction by type (e.g. central urban area, green urban area, countryside etc.) and region-specific density figures by type (ABF, 2006). Because Land Use Scanner simulates changes in residential land use and not in houses, the model is better fitted to simulate urban expansion than urban restructuring. For that reason, in the Second Sustainability Outlook study, housing construction in currently built-up areas was subtracted from the total regional housing construction target that was underlying the regional land claims for housing. In Tigris XL the number of houses to be constructed in currently built-up areas is contained in a separate input file at the zone level. Then, this number of houses in currently built-up areas is made consistent with the total regional housing targets, minus the housing developments realised at the new residential locations. In the scenarios, the share of brown field and green field developments differs depending on the strategy chosen for urban development.

Both the housing construction within and outside existing built-up areas is input for the housing market module in Tigris XL. The newly constructed houses are included as vacant houses by zone for the supply side of the module for choice of residential location. This module simulates the housing moves and location choices of the households. The module for choice of residential location and the demographic module together calculate for each year the changes in population (by age and gender) and households (by size and income class).

The labour market module in Tigris XL calculates the spatial distribution of employment for seven economic sectors, in the first step at a municipality level and in a second step at the level of the zones. The results (statistical regressions) of this module show that overall the spatial distribution of people influences the location choices of firms, although the existence and strength of this relationship differs strongly by economic sector (RAND Europe & Bureau Louter, 2006a). The changes in employment by sector, calculated by Tigris XL, are used as input for a post-processing step to calculate the regional claims for commercial and industrial land. This post-processing module was developed by Bureau Louter and for a description reference is made to Significance and Bureau Louter (2007). This step uses, besides the calculated changes in employment, information on location preferences of the sectors (e.g. for commercial sites), assumptions on the changes in these preferences in the future, and developments in average land use per employee per sector. These factors together determine the future land-use claims for commercial and industrial sites, public facilities and airports and sea ports. These land-use claims are calculated for the years 2020 and 2040 at a regional level, which is needed as input for Land Use Scanner.

In cases where *housing supply is market (demand) driven*, an alternative housing construction module is used in Tigris XL in which the housing construction rates at the level of zones are based on the location preferences of the household and the availability of land. To take better advantage of the detailed supply information in Land Use Scanner, the housing construction module in Tigris XL has been modified to be used in the integrated framework. In the new set-up, the main differences with the other approach are that demand surplus is now calculated specifically for each of the three residential area types and aggregated to the municipality level. Information on the highest suitability scores by residential area type in Land Use Scanner was used at the zone level to allocate housing construction within the municipality to the zones (see Fig. 5.3).

Under both market conditions, the Tigris XL model is used to generate the land-use claims for employment. The labour market module in Tigris XL calculates the changes in employment and a post-processing procedure module is used to convert these changes in employment into land-use claims for commercial and industrial land.

Land-use claims (in hectares) computed using output from Tigris XL at the COROP regional level were used as input for Land Use Scanner. Projected Tigris XL

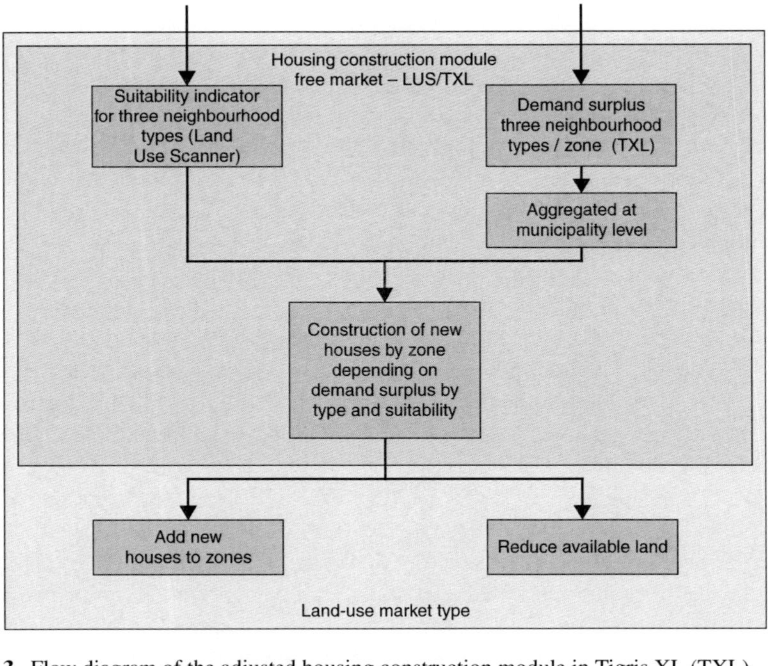

Fig. 5.3 Flow diagram of the adjusted housing construction module in Tigris XL (TXL)

demands for housing and employment by economic sector at LMS zone level were translated into land-use claims and subsequently used as input for the alternative land-use scenarios.

5.4 A Policy Application of the Combined Framework

The Land Use Scanner–Tigris XL framework was used in the Second Sustainability Outlook study to calculate the land-use and sustainability impacts of alternative scenarios and variants (MNP, 2007). The scenarios consisted of alternative future spatial and transport strategies for the Netherlands. One of the alternative land-use scenarios, the Uplands variant, is described here to illustrate the functioning of the combined framework.

The Uplands scenario entails a radical break with past trends in spatial development in the Netherlands. Under the scenario, new housing and employment areas, in the period from 2010 to 2040, are relocated from the low-lying, highly urbanised western part of the Netherlands (the Randstad area) to more peripheral areas lying above sea level and outside flood-prone areas near rivers. The Uplands scenario was developed to represent an extreme climate adaptation scenario. With current knowledge on the effects of climate change, the Netherlands is expected to be 'climate-proof' and protected from rising sea levels for some centuries to come. Structural spatial measures, such as a shift in investment to the upland areas of the Netherlands, are thus not urgently required. Yet, these were included in this scenario to show the potential implications of such drastic measures that are regularly suggested in adaptation studies.

To simulate the Uplands scenario, the housing construction module (see Fig. 5.3) used input data from both Tigris XL and Land Use Scanner. This scenario results in a markedly different land-use development compared to the Baseline scenario, which follows as much as possible current plans. Table 5.2 shows the differences in population development between the two scenarios; the country is divided into three parts - the urban core (Randstad), a surrounding intermediate area and the periphery in the far North, East and South of the country. The table shows that the strategy underlying the Uplands scenario results mainly in a shift in population from the urban core towards the higher, intermediate area. The periphery is also mainly located above sea level but this area does not attract many additional residents due

Table 5.2 Differences in population development according to the Uplands variant (in 2040) in different parts of The Netherlands

Part of the Netherlands	Population baseline scenario	Difference Uplands variant *minus* baseline scenario	Percentage population change in 2040
Randstad	7,826,331	−872,141	−11%
Intermediate zone	5,005,661	775,355	15%
Periphery	4,306,562	97,091	2%

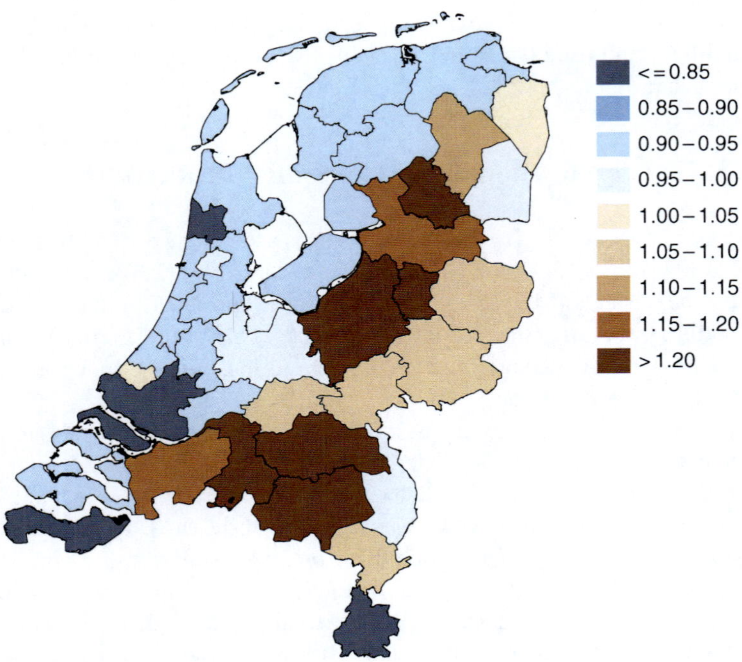

Fig. 5.4 Index of population development for Uplands scenario (2040 versus 2010)

to its relative poor accessibility to employment and services as compared to the Baseline scenario.

Figure 5.4 shows the population developments for the Uplands variant in the period 2010–2040 by region. The map shows clearly that the population in the area below sea level declines as a result of the assumed construction stop in combination with a falling average household size. Population growth takes place mainly in the regions above sea level and those bordering the urban core. Especially the province of Brabant, in the south of the country, can expect to experience high population growth. Areas in the middle of the Netherlands, where the river Rhine crosses the country from Germany to the North Sea, can expect lower population growth because the underlying strategy limits building options in this area due to increased risk of flooding. Although population development in the periphery is somewhat higher in the Uplands scenario than the Baseline scenario (see Table 5.2), there are regions in the periphery that nevertheless can expect a decline in population in the period 2010–2040.

Unlike residential developments, labour market developments are not restricted in the Uplands scenario. This assumption reflects Dutch planning practice, in which the construction of new houses is historically strongly regulated, while the location of employment is far less regulated due to oversupply of commercial real estate and competition between different regions. The large population shift from the Randstad to the intermediate zone is also reflected in the spatial growth pattern for employment, as presented in Fig. 5.5. The labour market module simulates

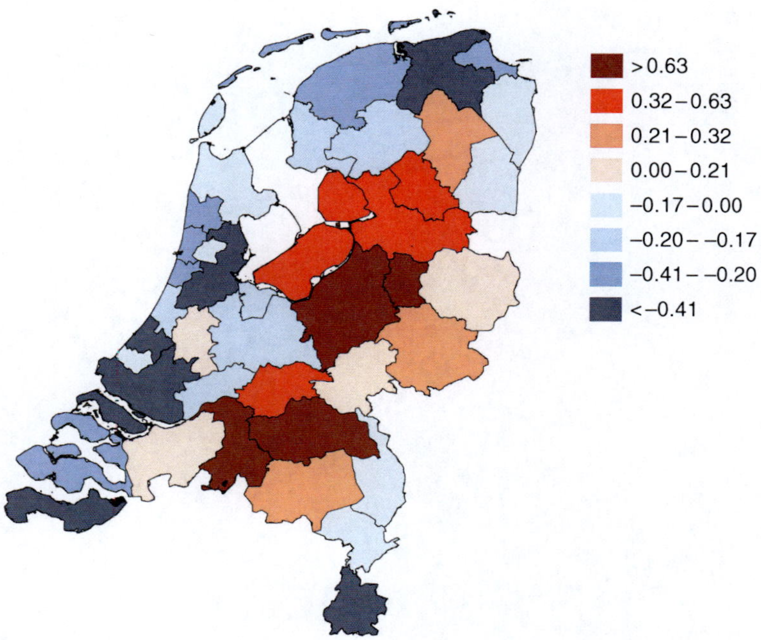

Fig. 5.5 Annual employment growth (%) by region under the Uplands scenario for the period 2010–2040

seven economic sectors and employment in most of these sectors, as consumer services or government, is influenced by population developments. Other sectors, such as commercial services, that respond to changes in employment are indirectly influenced by the population through changes in the employment resulting from the population following economic sectors.

Figure 5.6 presents the high-resolution land-use projections for the Uplands Scenario from Land Use Scanner, taking the population and employment developments at the COROP level from Tigris XL.

The integration of Land Use Scanner and Tigris XL makes it possible to calculate transport impacts and to include the mutual influences of changes in land use on transport, and vice versa. Different accessibility indicators were calculated referring to different definitions of accessibility. The congestion indicators show that the shift in population growth from the urban core to the intermediate area results in overall lower losses in traffic congestion time in 2040. Of course this pattern is regionally diverse, with a decline in congestion in the urban core and a strong increase in the intermediate zone. Applying a geographical accessibility indicator, representing the number of jobs that can be reached by car (during off-peak and peak hours) and public transport, shows a decline in the number of jobs that can be reached for the Uplands scenario as compared to the Baseline scenario.

As mentioned above, the Tigris XL model incorporates the LMS transport model of the Netherlands as its transport module. The LMS is a discrete choice type of transport model, which makes it possible to calculate the monetary accessibility

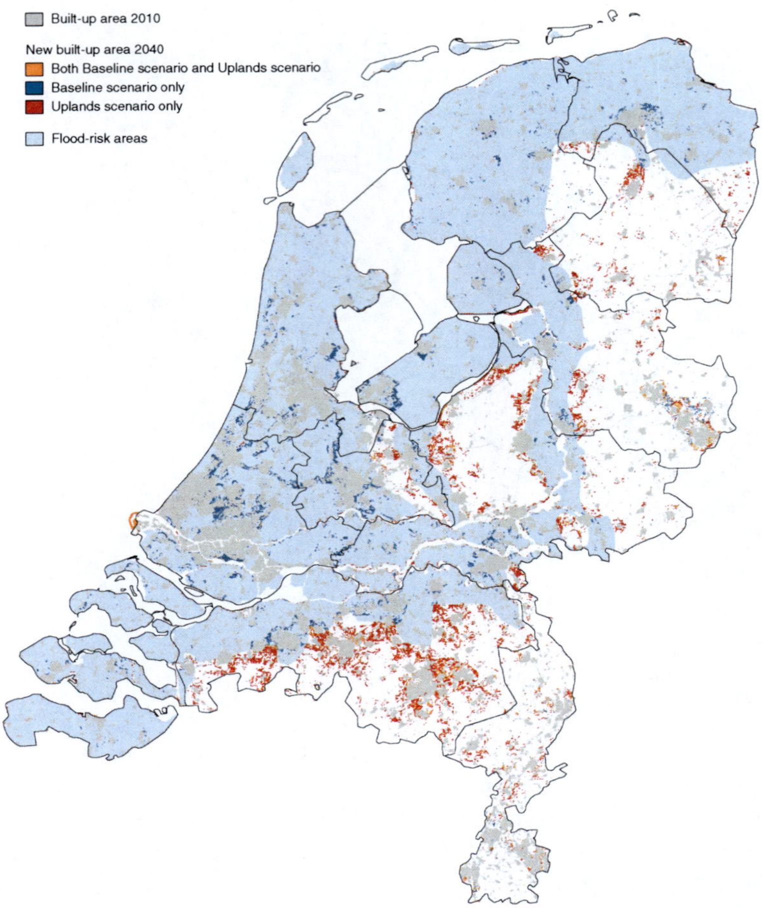

Fig. 5.6 Difference in land-use changes according to the Uplands scenario and the Baseline scenario, 2040

impacts following the logsum method (see Geurs, Zondag, De Jong & De Bok, 2010). The logsum method calculates, at a disaggregate level of person types, the differences in utility between a policy variant and a reference scenario. The cost coefficients in the utility functions can be used to convert the changes in utility into monetary terms. The advantage of the logsum method is that it can calculate the accessibility benefits of both transport and spatial strategies, while traditional measures only address the benefit of transport policies. The Uplands scenario does not assume any changes in the transport system and all accessibility benefits result from changes in the spatial planning strategy. Compared to the Baseline scenario, the Uplands variant would bring with it a high loss in accessibility totalling –521 million euros a year in 2020 (expressed in 2005 euros) and –1,182 million euros in 2040.

5.5 Conclusions and Directions for Further Research

This chapter describes the combined use of the Land Use Scanner and Tigris XL models within the same framework for the Second Sustainability Outlook study. The interaction between the models was set up pragmatically without making fundamental changes to either of the models. The combined modelling framework was used to analyse the impacts of alternative integrated spatial and transport strategies. Combining the modelling framework has a number of advantages:

- Detailed land-use data and plans from Land Use Scanner improve the simulation results of the residential location choices within Tigris XL and, thus, indirectly improve the simulation of the spatial distribution of employment (as these sectors influence each other).
- Conversely, Tigris XL simulation results for employment at a regional level, and associated changes in land use, can be used as land-use claims for Land Use Scanner in a feedback interaction. The Tigris XL model ensures consistency between population and labour developments.
- The combined Tigris XL and Land Use Scanner framework facilitates the analysis of the impacts of integrated land-use and transport policies. The results include impacts of changes in the transport system on the spatial distribution of residents/firms and land use and, conversely, the influence of changes in land use (and spatial distribution of residents and firms) on the transport system. Within Tigris XL, these mutual impacts can be analysed at the level of transport zones (1308 zones for the Netherlands); within the combined framework these effects are exchanged between Tigris XL and Land Use Scanner at the regional level. Land Use Scanner further allocates these changes to the grid cell level.

The inclusion of Tigris XL in the combined framework, and in so doing also the national transport model, enables the calculation of additional indicators, such as:

- Transport indicators for the alternative spatial and transport strategies, for example mode split and congestion effects.
- Accessibility impacts, expressed in geographical accessibility indicators as well as in the monetary benefits of changes in accessibility. The option to calculate the monetary benefits of changes in accessibility, due to transport policies as well as land-use policies, played an especially important role in the evaluation of alternative transport and spatial strategies. The results of the calculation showed that spatial strategies had large monetary accessibility benefits. This is, for example, illustrated by the much higher accessibility benefits of an urban density strategy, which was also considered in the sustainability outlook, in comparison with a large road-infrastructure investment programme (see Geurs et al., 2010).

The combined application of Land Use Scanner and Tigris XL was considered as promising and the framework has made a substantial contribution to the ex-ante evaluation of various policies, as discussed above. However, the interaction between the instruments was established only in a practical manner and a more fundamental integration of the instruments, or their respective knowledge, is still needed to further reduce overlaps and inconsistencies between the two and to fill in missing aspects.

A general observation is that simulations of the use of land (such as agriculture or residential), objects (such as houses or offices) and actors (such as people or firms) are closely related. Therefore an important direction for further research is to develop a more integrated and consistent approach towards modelling these three layers in time and space. It is clear that there are still many inconsistencies and omissions in the current combined framework when spatial changes in the number and distribution of residents, jobs, houses and in land use are calculated: inconsistencies and omissions between the three layers, between different spatial scale levels, and in time. To address these matters, we recommend several fundamental improvements of the framework:

- *Explicit modelling of objects and actors at a lower geographical level in integration with land-use changes.* Currently regional land-use claims derived from regional changes in population, housing or employment are allocated to the grid cell level. This often requires a post-processing step to transfer land use at the grid cell level into, for example, the number of people or firms. Information on the number of people or firms at the grid cell level is needed to be able to calculate indicators such as flood damage or environmental pollution. Explicit modelling of actors and objects within the region also enables improvements in the modelling of regional markets (e.g. housing and labour markets) and the behaviour of actors in these markets;
- *Use of one or more consistent theories throughout the analytical framework.* The use of a sequential set of models as described in this chapter, results in applying different theoretical principles at different spatial levels of scale. For example, different approaches are used for the simulation of choices of residential location within a region and for choices between regions. Although the underlying processes can be different at a regional and local level, for instance, more job-related migration at a regional level versus more housing-condition related moves at a local level, the use of a consistent approach (e.g. utility theory) would help to interpret modelling results in a consistent way in the policy evaluation.
- Improve the inclusion of supply information, such as land-use restrictions and options, in the modelling of regional developments. Ideally, this would mean that existing boundaries are less strictly enforced to allow conditions within a region to have an influence on interregional developments through feedback mechanisms (e.g. additional land-use restrictions in the Amsterdam region impact on the number of people in the nearby municipality of Almere). A first step has been made in this direction but additional steps are needed to improve the inclusion of land in the specification of the sector models.
- *Harmonise the temporal resolution of the model components.* TIGRIS XL is an incremental model that uses time steps of 1 year, while Land Use Scanner is

an equilibrium model that is normally applied for time horizons of 10, 20 or 30 years. Preferably, the interactions or data exchanges would take place annually to allow for path dependency and analysis of the developments over time.

References

ABF (2006). Scenario's en varianten Nederland Later, technical note prepared for Netherlands Environmental Assessment Agency, Bilthoven.

Geurs, K., Zondag, B., De Jong, G. C., & De Bok, M. A. (2010). Accessibility appraisal of land-use/transport policy strategies: More than just adding up travel time savings. *Transport Research Part D: Transport and Environment, 15*(7), 382–393.

Hilferink, M., & Rietveld, P. (1999). Land Use Scanner: An integrated model for long term projections of land use in urban and rural areas. *Journal of Geographical Information Systems, 1*(2), 155–177.

MNP (2007). *Nederland later. Tweede Duurzaamheidsverkenning*. Bilthoven: Netherlands Environmental Assessment Agency.

RAND Europe and Bureau Louter. (2006a). *Systeem documentatie TIGRIS XL v1.0, prepared for the Transport Research Centre*. The Netherlands: Leiden.

RAND Europe and Bureau Louter. (2006b). *TIGRIS XL toepassen in N18 studie*. The Netherlands: Leiden.

Significance and Bureau Louter. (2007). Toepassen van TIGRIS XL binnen de studie 'Nederland later'. Report prepared for Netherlands Environmental Assessment Agency, Bilthoven.

Significance and Bureau Louter. (2009). TXL analyse KBA RAAM. Report prepared for Netherlands Environmental Assessment Agency, The Hague, The Netherlands.

Significance, Bureau Louter and Stratelligence. (2007). Toepassen TIGRIS XL in de case studie indirecte effecten. Report prepared for Transport Research Centre, The Hague, The Netherlands.

Zondag, B. (2007). Joint modelling of land-use, transport and economy, T2007/4, TRAIL Thesis Series, The Netherlands.

Chapter 6
Biomass on Peat Soils?

Feasibility of Bioenergy Production Under a Climate Change Scenario

Tom Kuhlman, René Verburg, Janneke van Dijk, and Nga Phan-Drost

6.1 Introduction

Energy security has become a priority as the world's population increases and its standard of living improves, thus increasing energy consumption. As the demand for energy increases, there is growing concern about the possible exhaustion of finite supplies of fossil fuels in the not-too-distant future. In addition to the problem of availability, combustion of fossil fuels also has negative environmental effects: air pollution (e.g. particulates, nitrogen oxides, carbon monoxide and sulphur dioxide) produced through the combustion of fossil fuels, threatens human health as well as plant and animal life. Furthermore, the combustion of fossil fuels releases carbon dioxide and other greenhouse gases into the atmosphere, thus contributing to an increase in global temperature. These considerations lead to a search for alternative, renewable sources of energy, one of which is bioenergy.

Bioenergy, defined as energy from biomass, has the advantage that the carbon dioxide released into the atmosphere by its combustion is compensated for through the carbon dioxide removed from the atmosphere by the vegetation from which it was ultimately derived. Moreover, any country can produce bioenergy, so energy security is enhanced. There are two basic forms of bioenergy, that released by the simple combustion of biomass (e.g. firewood or animal dung), either directly for heating or indirectly for generating electricity; and that formed by the chemical transformation of biomass into other fuels (ethanol, diesel oil or biogas). One important form of the latter is the manufacture of liquid fuels from food crops such as vegetable oils (biodiesel), sugar and maize (ethanol). This form of bioenergy has become controversial because it competes with the use of crops for food, thus contributing to high food prices; it is also controversial because its environmental impact is considerable. Hence, there is doubt as to whether the use of these biofuels really mitigates the emission of greenhouse gases. Moreover, these biofuels

T. Kuhlman (✉)
Agricultural Economics Research Institute (LEI), PO Box 29703, 7502 LS The Hague, The Netherlands
e-mail: tom.kuhlman@wur.nl

E. Koomen, J. Borsboom-van Beurden (eds.), *Land-Use Modelling in Planning Practice*, The GeoJournal Library 101, DOI 10.1007/978-94-007-1822-7_6,

are costly, which means that market-distorting policies are needed to promote their use.

There is, therefore, a need to find more efficient ways of producing bioenergy. Some hope has been placed on so-called second-generation biofuels. These encompass a wider spectrum of materials, notably crop waste (such as straw) and plants that can be easily grown but are not suitable for food (e.g. many grasses). The most promising types of second-generation biofuels use woody and fibrous biomass (lignocellulose) rather than sugars and oils. The technology for breaking down lignocellulose into the sugars from which ethanol can be produced exists, but is not yet economically attractive. There are hopes that further technical progress will lead to efficient large-scale production of such fuels in the near future. This would mean establishing plantations of the appropriate crops, thus leading to the question of where these could be planted.

To explore that question for the Netherlands we have opted for two types of wetland crops: reed and willow. In a densely populated country like the Netherlands, the feasibility of growing a low-value crop on a large scale is not immediately apparent. However, in the future, conditions are likely to be different from those of today. For one thing, climate change and sea level rise may result in some current uses of agricultural land becoming uneconomical. For another, multifunctional land use may provide a way out: biofuel need not be the sole product of the land. In the case of the Netherlands, using wetlands to grow these crops should provide additional benefits. Firstly, the land can be also used as a flooding buffer – the need for which may increase as a consequence of sea level rise and increased volumes of river flow. Secondly, since reed and willow are indigenous to low-lying areas in the Netherlands, their cultivation could enhance Nature values – albeit with some limitations. Their cultivation would help create naturally aesthetic wildlife habitats. The harvesting of the biomass would create a disturbance, of course, but this could be mitigated by appropriate timing of the harvest. Thirdly, both types of vegetation can have beneficial effects for land and water resources. Reed is an effective water-filtering agent and willow has the capacity to remove certain metals from the soil. Indeed, willow plantations have proven to be effective in soil remediation programmes and are being used commercially for this purpose (Bertholdsson, 2001; Greger & Landberg, 1999). Willow has been used experimentally in natural filtration beds in a manner similar to reed beds for the filtration and removal of nutrients from of waste water (biofiltration) and in buffer zones to protect sensitive areas from nutrient leaching (Larsson et al., 2003). All these benefits can be added to those of biomass production itself.

Will multifunctional use be enough to make this type of land use feasible and, if so, where are suitable areas for growing this biomass to be found? That question is explored in this chapter, which looks specifically at the economic feasibility of these biofuel crops under present conditions, how these conditions might change, and what kind of change would be necessary to make growing these biofuel crops feasible.

6.2 Method

To identify areas where suitable conditions for industrial-scale biomass production might exist, use was made of the spatial model Land Use Scanner, a GIS-based model for generating spatial forecasts of land use. The model allocates land to various classes of use according to those types of land use that generate the largest benefits for society. It is, therefore, based on economic theory: Land Use Scanner simulates competition between land-use classes in which the benefits are expressed as bid prices for land (Hilferink & Rietveld, 1999).

Land Use Scanner is appropriate for our study because it considers not only the suitability of land for biomass production, but also the claims of all other types of land use: residential, industrial, agricultural, nature, and water management. Claims on land and suitability for each land-use class are dependent on the particular scenario under consideration. The suitability of each location (the model uses a large number of grid cells) is influenced by current land use, physical properties (soil type, groundwater), operative policies (zoning, water management, protection of nature areas), proximity to other land uses and accessibility (Koomen, Loonen & Hilferink, 2008). These characteristics are included in the database accompanying the model. However, the values attached to each of these aspects are governed by the particular scenario chosen for a particular model run.

For our study, we designed two scenarios and compared their results: one scenario is an extrapolation of recent trends – called the *reference* scenario; the other is a constructed scenario within which crucial factors affecting the feasibility of biofuel production are varied so as to see under which conditions this particular type of production might become feasible – called the *bioenergy feasibility scenario*. Both scenarios incorporate socio-economic factors and the impact of climate change as driving forces of land-use change: they both use the socio-economic scenarios described in a study published by three Dutch planning agencies in 2006 (CPB, MNP & RPB, 2006), and the climate scenarios published in 2006 by the Royal Netherlands Meteorological Institute (Van den Hurk et al., 2006). The particular socio-economic scenario used in both our scenarios is the 'A1' version. This scenario is also referred to as 'Global Economy' and assumes market liberalisation and high rates of economic and demographic growth. The biomass feasibility scenario assumes strong climate change and introduces a high energy price as an additional driver. The reference scenario is presented for the sake of comparison.

In our study, the Land Use Scanner model used 100 m × 100 m grid cells, offering a high resolution of land use. All land-use classes were simulated in Land Use Scanner's new, discrete allocation algorithm, which allocates only one type of land use to each cell; for more details, see Chapter 1. Since land-use claims are one of the basic scenario-dependent inputs of the model, claims for industrial-scale biomass production need to be established. This claim-set was evaluated by cost-benefit analyses of biomass production from reed and willow under a variety of conditions.

To use Land Use Scanner for this study we introduced a new land-use class for biomass production. This is a multifunctional class, reflecting the additional benefits that this type of land use could generate: apart from biomass production, it includes water storage with some water filtration services and nature enhancement added. Since land is allocated to a particular land-use class based on the net benefit to be obtained from that land use, we first examined what these benefits were. This entails a cost-benefit analysis, which is dealt with in the next section.

6.3 Cost-Benefit Analysis of Biomass Production

Cost-benefit analysis (CBA) is normally used to assess whether a project is worthwhile, in comparison to alternative uses of the investment financing required. In our case, the question is whether a particular type of land use delivers greater benefits than the alternatives. Costs and benefits can be seen in social terms, i.e. the costs and benefits of land use to society. Private CBA, on the other hand, addresses the question of whether it will be worthwhile for the individual land user to allocate his or her land to the use in question.

Our focus here is on social CBA, because we need to know under what conditions it will be advantageous to allocate land to biomass production. We compared the gross value of agricultural production of two biofuel crops with that of likely alternative agricultural uses, while also assessing the value of additional benefits to be obtained from using the land for reed and/or willow production. Considering private CBA would require that we look at net returns to the farmer, i.e. just incorporating production costs; we only look at processing costs after the produce leaves the farm.

In temperate climates, reed (*Phragmites australis*) and willow (*Salix* spp.) grow naturally in swampy areas. On well-drained soils, providing the groundwater is maintained at present levels, reed can produce 35 tonnes of dry biomass per hectare while willow can produce 15 t/ha. In reed, 45% of this biomass is cellulose, whereas in willow the proportion is 37.5%. As each tonne of cellulose can generate 340 L of ethanol, 1 ha of reed can therefore produce 5,355 L of ethanol, while 1 ha of willow generates 1912.5 L. These are equivalent to 4.23 and 1.51 tonnes of ethanol, respectively, at a specific gravity of ethanol of 0.789. From that we can estimate the energy generated per hectare of biomass, since one tonne of ethanol yields 26.7 gigajoules (GJ). Thus, 1 ha of reed generates 112.9 GJ and that of willow 40.3 GJ. Much higher energy values can, however, be obtained from direct combustion of dry matter, for heating or for the generation of electricity: 620.0 GJ/ha of reed and 274.89 GJ/ha of willow. Reed and willow have near maximal production values in conditions that are too wet for grass or field crops. The yield of reed in waterlogged soils is only 1.8% lower than that in well-drained soils (Hellings & Gallagher, 1992).

The next question is: what return can a farmer expect? This requires us to take a look at energy prices and their possible evolution in coming decades. World population growth together with economic growth will lead to a large increase in the demand for energy up to 2050. Energy supply will, however, have difficulty in

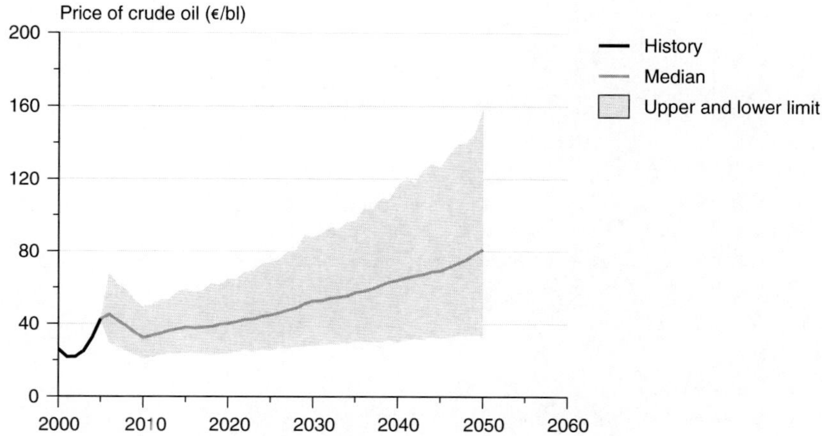

Fig. 6.1 Projected prices (€/barrel) of crude oil up to 2050

meeting this increased demand, as oil and gas reserves are expected to decline and remaining reserves will be more expensive to exploit. Although there are alternative sources of energy, these tend to be expensive. All this points to a future in which rising energy prices will make bioenergy economically more attractive. Projections of future oil prices were made with the PROMETHEUS model to calculate the prices of crude oil per barrel up to 2050 (Fig. 6.1). We have used the upper limit for the bioenergy feasibility scenario (i.e. a price of €160/barrel) and the median projection for the reference scenario (€80/barrel).

The cost price of harvesting, transportation and processing of biomass into ethanol in 2005 was US$ 60/t, which is equivalent to €50 at the average exchange rate in 2004/2005. Allowing for technological advances in processing, we assume this cost to drop by 1%/year, which implies a cost price of €31.5/t in 2050. Direct combustion requires less cost than fermentation: US$ 50, or €41.7, per tonne in 2005. By 2050 this could be as low as €26/t. This enables us to compute the net private benefit of biomass production. Figures 6.2 and 6.3 show the results for conversion of biomass into ethanol: Fig. 6.2 for the reference scenario, under which groundwater levels are maintained as at present and oil prices follow the median prediction; and Fig. 6.3 for the bioenergy feasibility scenario, with higher levels of groundwater and oil prices at the upper limit.

These calculations show that ethanol production based on reed can provide a return to farmers if oil prices rise to levels such as those of 2008, before the current financial crisis caused a steep decline. It could become profitable in 20 years or so, even with lower oil prices, provided biomass processing technologies progress at a reasonable pace. Ethanol based on willow would be marginally feasible only under the most favourable conditions. Biomass production for direct combustion process offers more profit to farmers. The net benefits of both reed and willow are all positive, even for modest oil prices (Figs. 6.4 and 6.5).

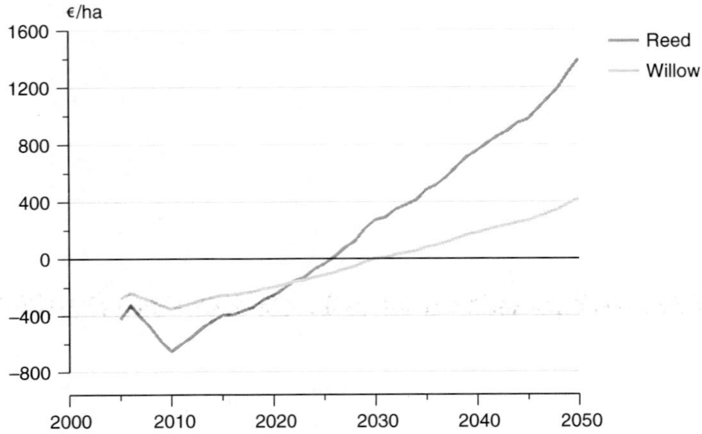

Fig. 6.2 Net benefit of biomass conversion to ethanol for the reference scenario

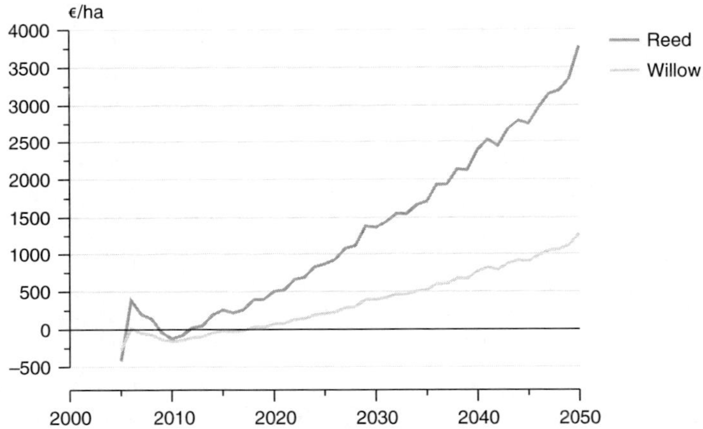

Fig. 6.3 Net benefit of biomass conversion to ethanol for the bioenergy feasibility scenario

These figures are gross potential returns to farmers. How competitive can reed and willow be, given the present uses of farmland? An economic study of willow cultivation in Northern Ireland shows that, with gross margins excluding subsidies for suckler cows and for lowland sheep, willow coppicing can compete with grassland use, depending on circumstances on each individual farm (Rosenqvist & Dawson, 2005). In the Netherlands, potato is the most important field crop and grass for cattle the most important alternative land use for larger areas (i.e. excluding horticulture and intensive livestock-keeping). Based on average yield data for the period 2005–2007, we calculated the value added per hectare per year that these two uses of land bring the farmer. Potatoes can be grown only once every 4 years; less profitable crops have to be grown in the other 3 years in order to keep the soil healthy, which has been taken into consideration in the calculation of the

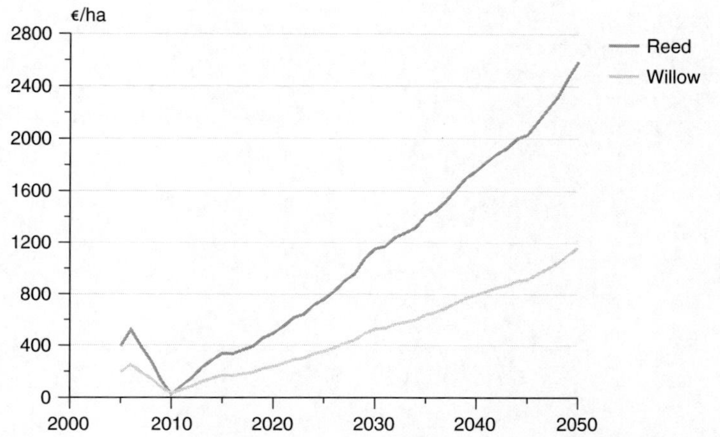

Fig. 6.4 Net benefit of biomass combustion for the reference scenario (median oil price)

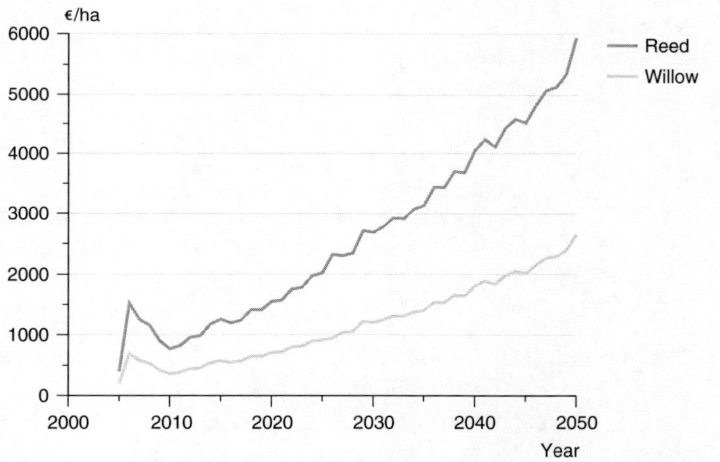

Fig. 6.5 Net benefit of biomass combustion for the bioenergy feasibility scenario (high oil price)

average value added per year. One hectare of arable land used in this way yields roughly €9,000 for potatoes, €2,400 for sugar beet and €800 for cereals (LEI, 2008). If it is assumed that cereals are grown for two of the 3 years, the annual return for potato is €3,200/ha. Grassland used for dairying generates approximately €4,800/ha (Table 6.1).

These value-added figures must be compared with those for bioenergy production (shown in Table 6.2). We see that bioenergy can compete only if processed through direct combustion, and then only at fairly high oil prices. This is, however, only for the current situation. As Figs. 6.2, 6.3, 6.4, and 6.5 show, value added for bioenergy may rise in decades to come. In our simulation, energy yields of reed and willow were kept constant; although in reality they may rise, the same is likely to happen for other agricultural yields, so the competitive positions of different types of land

Table 6.1 Average value added for the products associated with the two major farming systems in the Netherlands (2005–2007)

Product	Area (× 1,000 ha)	Value added (€/ha)
Milk (per ha grass and fodder maize)	1,209	4,809
Winter wheat	209	1,025
Summer wheat	120	785
Summer barley	45	690
Sugar beet	85	2,440
Potatoes (consumption)	69	8,959
Potatoes (starch)	49	2,494
Grass seed	25	1,314

Source: LEI (2008)

Table 6.2 Potential value added of two different types of bioenergy production in 2005–2007 at current groundwater levels and two different oil prices

Bioenergy type	Oil price = USD 58/barrel[a]	Oil price = USD 85/barrel[a]
Conversion to ethanol		
Reed	€-431/ha	€199 /ha
Willow	€-278 /ha	€/-50 ha
Direct combustion		
Reed	€391 /ha	€1249 /ha
Willow	€194 /ha	€575 /ha

[a]These are the median and upper limit projections, respectively, for oil prices generated by the PROMETHEUS model for 2007. Actual oil prices in that year varied from $48 to 91/barrel (world average spot-market prices)

use will not be affected. Only for the cost of processing biomass have we assumed a gradual reduction. Increasing liberalisation is, however, likely to have a negative effect on prices of milk and potatoes. Furthermore, if climate change leads to rising groundwater levels (as we have simulated in one of our scenarios), the yield of grassland and of potatoes will be reduced, whereas reed and willow will be little affected. These changes may make biomass production attractive in some areas, as is explored in Section 6.5.

However, bioenergy would be only one source of economic return from the land under the proposed form of land use. Other returns/benefits include:

- Surface water storage: reed and willow grow well on low-lying land and do better than other crops under waterlogged conditions. This makes them suitable crops for areas reserved for water retention, such as areas that may be temporarily flooded to counter excessive water levels in rivers or other water bodies, or to store fresh water to cope with drought and to combat salinity.
- Natural water filtration: reed acts as a natural filter for purifying water. Moreover, waste water or sludge could stimulate wood biomass production substantially and the nutrients it contains could replace conventional fertilisers to a large extent. Research by Larsson et al. in 2003 has concluded that growth levels of willow

after a first 3-year rotation were higher under irrigation with waste water than for non-irrigated cultures or those irrigated with drinking water.
- Nature values: since reed and willow are indigenous plants in wetter parts of the Netherlands, their cultivation can enhance natural attributes and provide a habitat for wildlife, thus stimulating the value of these areas for recreational activities and local tourism.

To some extent, these benefits compete with one another and it will not be possible to optimise all of them simultaneously. Using waste water to fertilise a biomass crop will subtract from its nature value. Similarly, harvesting the biomass will disturb wildlife, although appropriate timing could minimise some of its negative effects. In this study, we have refrained from quantifying the filtration and nature benefits of this form of land use, confining ourselves to just mentioning them as additional potential benefits.

Concerning water storage, it is the policy of the Dutch government to cope with climate change by providing space for excess water, rather than, as in the past, just raising the dikes and pumping the water out (Tielrooij, 2000). The current slogan is 'retain, store, discharge' in that order. In the strategy of 'Room for the River' (VROM, 2006), forty locations in the Netherlands have been identified as sites for retention buffers. The existing dike around the polder is to be breached to allow the river to flow onto the low-lying land at high water. Although this does not preclude agricultural use of such land, no high-value uses such as building or cultivation of perennial crops will be permitted – thus reducing the value of the land. If it is farmland (which will usually be the case), the farmer will have to be compensated for this loss of value. The economic benefit of water storage has been estimated to be €395/ha, or 4 eurocents/m^2/year, as this is what the Dutch Ministry of Agriculture will pay as compensation to farmers whose land is used for water retention (Verburg & Jongeneel, 2008).

6.4 The Purpose of the Scenario Approach

To assess the potential for bioenergy production in the Netherlands, two scenarios have been drawn up that attach differing importance to several crucial conditions for biomass production. One scenario – the reference scenario – assumes a continuation of currently prevailing conditions; whereas the other – the bioenergy feasibility scenario – assumes conditions that make biomass production feasible. We, thus, set out to examine to what extent the feasibility of this type of land use depends on these favourable conditions. The differences between the two scenarios relate to:

- demand for land from outside the agricultural sector (in turn, determined by population dynamics and economic growth);
- energy prices prevailing on the world market;
- the policy environment in relation to European agriculture;
- the extent and speed of climate change;
- government policies for adapting to climate change.

6.4.1 Reference Scenario

In this scenario, the assumptions concerning economic and demographic growth are in line with the Global Economy scenario in the Dutch study 'Welfare, prosperity and quality of living environment' (CPB et al., 2006). This scenario is also applied in preceding climate change adaptation studies in the Netherlands (see, for example, Koomen & Van der Hoeven, 2008). The main scenario assumptions and their application in the Land Use Scanner model have been documented extensively elsewhere (Koomen et al., 2008; Riedijk, Van Wilgenburg, Koomen & Borsboom-van Beurden, 2007). In short, the following assumptions apply:

- The Dutch population will grow to around 20 million by the year 2050. Average household size will decrease further and economic growth will be substantial. This will raise demand for accommodation for living, working and recreation. As a consequence, the area of land available for agricultural will decline rapidly. These assumptions follow
- The median projection for oil price will apply, leading to a price of €81/barrel (in 2004euros), as compared to €160 projected for the upper limit, by the year 2050;
- Protection and subsidies for European agricultural products will be withdrawn and European farms will have to operate in a liberalised market environment. Producer prices will be lower and a large part of the food consumed by Europeans will be imported from countries that can produce it more cheaply.
- Climate change is assumed to be minor in our reference scenario, following the moderate (G) scenario, as described by the Royal Netherlands Meteorological Institute (Van den Hurk et al., 2006). Table 6.3 provides more details of the climate change scenarios.

Table 6.3 Key parameters of two Dutch climate-change scenarios for 2050 relative to 1990 values, often referred to as G (moderate) and W+ (warm scenario with changed circulation patterns)

Climate parameter	G	W+
Winter		
Mean temperature	+0.9°*C*	+2.3°C
Annual coldest day	+1.0°C	+2.9°C
Mean precipitation	+3.6%	+14.2%
Summer		
Mean temperature	+0.9°C	+2.8°C
Yearly warmest day	+1.0°C	+3.8°C
Mean precipitation	+2.8%	−19.0%
Potential evaporation	+3.4%	+15.2%
Annual values		
Annual maximum daily mean wind-speed	0%	+4%
Absolute sea level rise (cm)	15–25	20–35

Source: Van den Hurk et al. (2006)

- Water policies will be aimed at controlling flood threats by conventional means: strengthening and enlarging dikes; sand suppletion against coastal erosion; limited additional space for rivers;
- As a consequence of these policies, in combination with modest changes in climate and hence water levels, groundwater levels are maintained at the same levels as today.

6.4.2 Bioenergy Feasibility Scenario

The bioenergy feasibility scenario has been designed in such a way that its principal driving forces point towards conditions that are favourable for biomass production in combination with water storage. In other words, this scenario examines whether bioenergy production could be feasible under favourable conditions and, if so, the likely location of that production. In general, the driving forces are the same as in the reference scenario. However, the following special conditions apply that are favourable to biomass production in combination with water storage:

- Energy prices will follow the upper limits of the projection shown in Fig. 6.1. This is consistent with high economic growth (globally, as well as in the Netherlands).
- Climate change is assumed to follow the relatively warm scenario with changed circulation patterns (W+), see Table 6.3. The climate change scenario selected is characterised by a temperature rise of 2°C between 1990 and 2050, plus concomitant change in atmospheric circulation patterns. Winters will be less cold but wetter because of an increase of westerly winds; summers will be drier and warmer because of prevailing easterly winds (Van den Hurk et al., 2006).
- Protection against flooding will be provided partly by 'returning the land to the water', for example, through the designation of water retention areas. Higher water levels in both the sea and rivers (particularly in winter) are also expected in this scenario. Combined with the new water-retention policy, this will lead to higher groundwater levels and occasional inundation of low-lying areas – at least in those areas that can be flooded without extensive economic damage.

6.5 Implementation and Results

Our next step was to assess in a spatial dimension the potential for biomass production in relation to other agricultural crops. This was done by applying Land Use Scanner (described in Section 6.2), using the relative benefits of the different land-use classes (described in Section 6.3), to the scenarios drafted Section 6.4. Modelling with the Land Use Scanner involves the construction of suitability maps and the specification of a (regional) demand for land, which are reported in this section together with the results of the simulation. Here we limit ourselves to specific implementation issues related to biomass production, as the basic implementation of the Global Economy scenario is described extensively elsewhere (Riedijk et al., 2007).

6.5.1 Suitability Maps

The physical or spatial characteristics of locations, in our case 100 m × 100 m grid cells, are described in the suitability maps with scenario-dependent values relating to these characteristics. The local suitability for the bioenergy land-use types is determined by the potential values of biomass production and water retention.

The potential value of biomass production is based on the cost-benefit analysis introduced in Section 6.3. The yield values presented there, however, only apply for optimum conditions. Therefore, a yield-loss fraction is specified for each location, determined by soil type and hydrological conditions. These yield-loss fractions have values between 0 and 1, with zero yield-loss representing optimum growth conditions. These calculations were also made for the principal competing land uses, potatoes and grassland.

To describe local soil types, a simplified soil map of the Netherlands was constructed based on the soil map 1:250,000 (WUR-Alterra, 2006), distinguishing between clay, sand, loam and peat (organic) soils. Groundwater data were derived from the groundwater map, which has a resolution of 100 m × 100 m (WUR-Alterra, 2006).

Martin and Stephens (2006) report in their study that willow grows well in coarse to fine loamy soil textures of alluvial or fluvial deposits. In consequence, clay soil and sand soil have a higher yield-loss fraction than other soil types. We used the following yield-loss fractions for willow: loam = 0.0; peat = 0.1; partially organic soils (with a lower percentage of organic material, and the rest made up mostly of sand) = 0.2; sand = 0.4; clay = 0.6.

Reed grows in marshes and swamps, in firm mineral clays, where groundwater levels fluctuate from 15 cm below soil surface to 15 cm above – as is the case along streams, lakes, ponds, ditches, and wet wastelands; reed tolerates moderate salinity (Duke, 1983). Most soil types in the Netherlands are suitable for growing reed; only sandy soils have a somewhat lower potential, and thus a higher yield-loss. Reed yield-loss fractions are: clay = 0.0; loam = 0.2; peat = 0.2; partially organic soil = 0.2; sand = 0.3.

These soil-based yield-loss fractions are, of course, independent of the scenario. As was discussed previously, this is not the case for groundwater levels. Thus, separate suitability maps per crop are needed for each scenario. The bioenergy feasibility scenario does, however, not only have higher groundwater levels than the reference scenario, but also higher energy prices. The net effect is much higher returns than under the reference scenario.

Before the suitability maps could be set up, there was one more step to be taken. Since Land Use Scanner simulates a land market in which land-use classes act as rival bidders for land, the suitability has to be expressed as a bid price reflecting the value of that particular location for that particular land use. Therefore we had to convert yield figures into bid prices. This was done by calculating the present value of the land for the particular crop: revenue per year was added up for a number of years and discounted for each year, reflecting the fact that a profit in the future is less valuable than the same profit today. The common discount rate for public investment

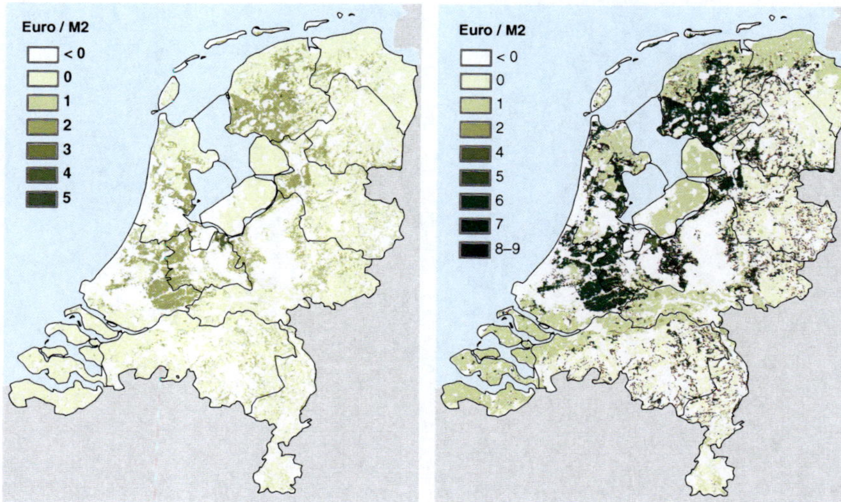

Fig. 6.6 Suitability maps for reed in 2050 based on the median oil price projection and normal groundwater levels (*left*); and on the upper limit (*high*) projection of oil prices and high groundwater levels (*right*)

projects in the Netherlands is 3.5%.[1] Over a period of 20 years, this results in a land value of about 15 times the annual yield. This is equivalent to a bid price in 2050 of €3.8–€8.5/m^2 for the cultivation of reed, depending on the oil price (assuming that the reed will be used for combustion rather than for distilling ethanol). For willow, the bid prices are, of course, much lower: from €1.7 to €3.9/m^2. The resulting suitability maps for reed are shown in Fig. 6.6.

The areas suitable for reed are primarily low-lying peat soils and partially organic soils. Being low-lying, they are also suitable for water storage. Wet conditions mean yield loss, but for reed this loss is less than for field crops or grass, which means that more areas will be suitable for reed under such conditions, as the Fig. 6.6 shows. For willow, there is hardly any loss at all, as was discussed earlier. Figure 6.7 shows the suitability maps for willow. The pattern is very similar to that of reed; the differences are due to a lower bid price for willow and to slightly different soil requirements.

These bid prices can be compared to those of potato (€4.7, calculated in the same way as above) and dairy production (€7.0). Potatoes are grown on relatively

[1] The appropriate discount rate is controversial, as is the period over which discounting should be done. However, for comparing the competitive attractiveness of different agricultural uses it makes no difference what method is used, as long as it is the same method for all. What is important is that a bid price is reached that realistically reflects the competitive strength of the agricultural sector as compared to other land uses. This strength is low: almost all other land uses offer higher returns per hectare than agriculture, which means that agriculture effectively functions as the balancing entry in the land use accounts. The only land uses offering lower returns are for forest and Nature, but these tend to be protected by law. The values arrived at here for agricultural bid prices remain well below those used for competing land uses in Land Use Scanner (e.g. €20/m^2 for recreation).

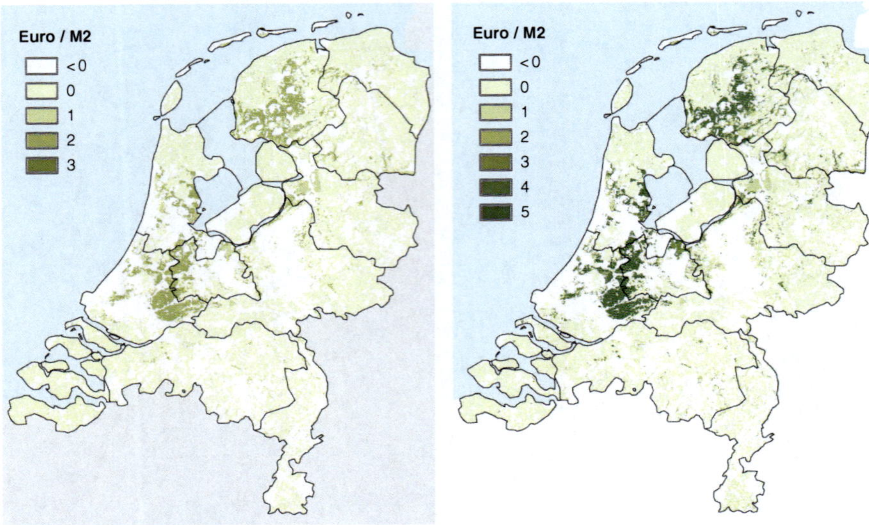

Fig. 6.7 Suitability maps for willow in 2050 based on the median oil price projection and normal groundwater levels (*left*); and on the upper limit (*high*) projection of oil prices and high groundwater levels (*right*)

dry land, but dairy production has a suitability pattern fairly similar to that of reed and willow, as Fig. 6.8 shows. The difference is that, since the value added per unit area of dairy production under the reference scenario is higher than for the bioenergy feasibility scenario, the potential area for dairying is much larger.

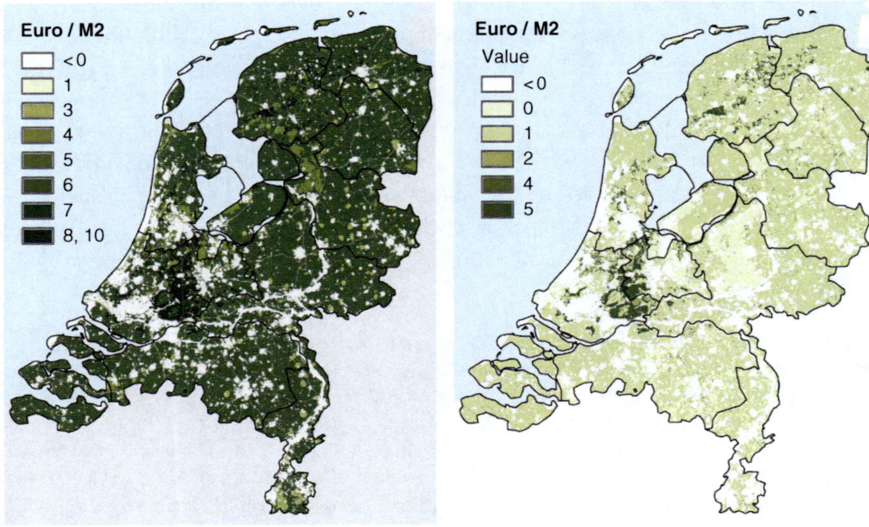

Fig. 6.8 Suitability maps for dairy production in 2050 based on the median oil price projection and normal groundwater levels (*left*); and on the upper limit (*high*) projection of oil prices and high groundwater levels (*right*)

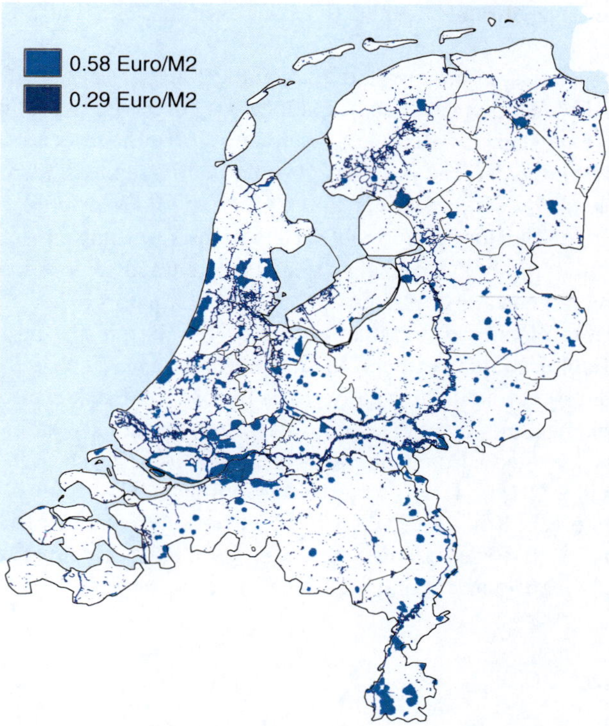

Fig. 6.9 Additional benefit (increased bid price) for biomass suitability reflecting the local potential for water storage (2050)

With high groundwater levels and without market protection, however, the dairy yield is reduced, as the right-hand map demonstrates. Yield reduction due to wetter conditions is much less for reed and willow and, moreover, these crops do better because of higher energy prices.

A final element in the suitability map definition for the bioenergy land-use types is the expected additional benefit of water storage (see Section 6.3). This aspect is based on elevation – naturally, low-lying areas are needed; and hydrological status – areas adjacent to rivers, canals, lakes and already identified retention areas are categorised in different degrees of suitability. This means that the local suitability value (bid price) can be augmented with this additional benefit. Using the value of €395/ha/year presented in Section 6.3 and the bid price calculation method described above, we arrived at an increase in bid price of €0.58/m^2. The spatial pattern of this additional benefit, which is the same for both scenarios, is shown in Fig. 6.9.

6.5.2 Land Claims

Apart from mapping the suitability of each particular cell for each type of land use (using a maximum bid price as a standard from which each cell can deviate), the Land Use Scanner model also requires as input a total claim for each land use, that is, the total amount of land required for that land-use class. This, too, is scenario dependent.

The model allows minimum and maximum claims, i.e. the land minimally required and the maximum that can be allocated. For biomass production, we set the minimum at 0, meaning that the model is not forced to allocate any land to it. At the maximum, the demand for energy would be high enough to cover the entire country with reed and willow. This is unlikely, of course. For example, people attach a high value to the historical agricultural landscapes existing in the Netherlands, some of which are unique. Covering these with reed and willow would make the landscape monotonous and obliterate a thousand-year history. Therefore, we set the maximum extent of biomass production at 15% of the present agricultural area. Since the agricultural area is bound to shrink in the period up to 2050, it will be more than that proportion in the target year. A better-founded demand for bioenergy in the Netherlands, taking into account international market conditions, could be obtained from specified agro-economic models such as LEITAP (Banse, Van Meijl, Tabeau & Woltjer, 2008). This is foreseen for future studies.

For water storage, Kok (2004) calculated a need of 93,678 ha of water retention and storage by 2050. However, part of this consists of existing primary watercourses, which leaves 72,226 ha needed for additional water storage.

6.5.3 Results

The results for both scenarios are shown in Fig. 6.10. The model allocates very little land (3,350 ha, to be exact) to biomass crops under the reference scenario. In other words, under current hydrological conditions and with oil prices at a level of about €80/barrel (or about $100–110 at the exchange rate prevailing in the first half of 2009), biomass cultures on wetlands are unlikely to be an attractive option for farmers except on a few locations. However, climate change and measures to adapt to it could alter hydrological conditions on low-lying farmland (especially peat soils now used for grassland) so as to make these areas less suitable for pasture and hay and more suitable for reed and willow. If this is combined with higher oil prices (at a level similar to or higher than those prevailing in the first half of 2008), reed would become commercially interesting in large parts of the Netherlands. Land Use Scanner allocates to it 339,000 ha, which is almost 15% of all agricultural land in use today.

Willow gives lower yields than reed, so from the point of view of energy production the scenario would lead to the cultivation of reed only. However, mixing reed with willow stands would make for a less monotonous landscape, albeit with slightly lower biomass production.

The areas allocated to biomass production in the map on the right-hand side have a variety of soils – clay, sand and peat; but the larger contiguous areas are peat soils. These are also the areas for which the long-term sustainability of dairy production is questionable, since optimal growth of grass requires lowering the groundwater to a level at which oxidation and subsidence of these vulnerable soils becomes inevitable

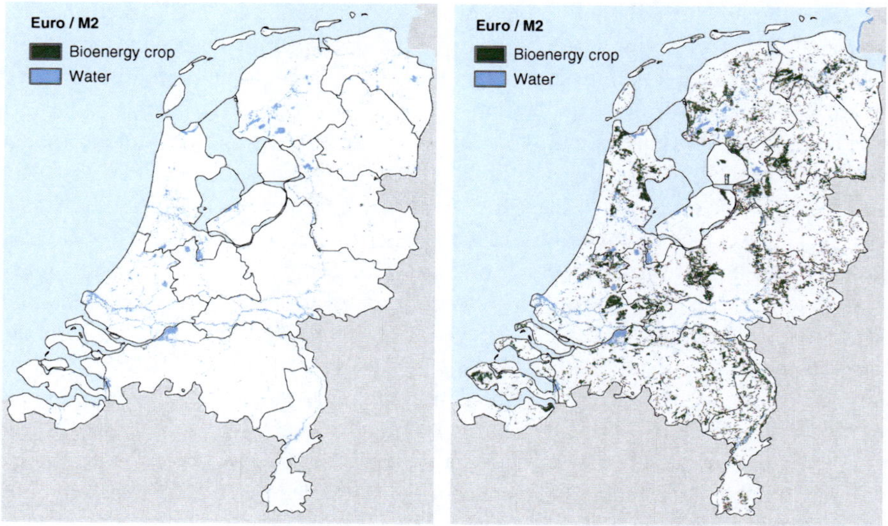

Fig. 6.10 Land use for reed with water storage in 2050 under two scenarios: for the reference scenario (*left*); and for the upper limit (*high*) projection of oil prices and high groundwater levels (*right*)

(Van den Akker, 2005). Crops such as reed and willow that can be grown under wet conditions may well be the salvation of these soils.

6.6 Conclusion

Reed (*Phragmites australis*) and willow (*Salix* spp.) are potential feedstocks for bioenergy. Their biomass can be processed either into cellulose-based ethanol or burned directly to generate electricity. Since the first process itself requires much energy, the second appears to be the most efficient manner for harnessing this energy source – at least with current technologies.

Both plants are indigenous to the Netherlands, and do well under wet conditions. This makes them particularly suitable for the peaty soils in low-lying areas. These areas are presently used mostly for pasture (dairy cows and sheep), and we have, therefore, examined the economic viability of reed and willow in relation to grassland pastures for dairy production. This viability is also given a spatial dimension: we investigated *where* growing bioenergy crops might be advantageous.

These bioenergy crops offer other potential benefits as well. The land used for these crops can also be used for temporary water storage, either seasonally or as a reserve in case of calamity. The need for such areas is likely to increase as climate change progresses – albeit depending on whether a policy of adaptation to increased water levels is opted for or one that raises dikes and increases the volume of water to be pumped out. Furthermore, this type of land use may also improve Nature values, although this will depend on how the land is managed. Finally, reed also has a propensity for purifying water. A possible disadvantage of this scenario could be the creation of a monotonous landscape, although this could be partially alleviated by mixing reed and willow production.

Quantifying only the benefits of energy production and water storage, under currently prevailing conditions (2009) these wetland bioenergy crops offer significantly lower benefits than for grassland agriculture. Conditions change, however, if we factor in, on the one hand, the impact of climate change (higher groundwater levels as a consequence of sea level rise) and, on the other hand, higher energy prices (a consequence of increasing scarcity of fossil fuels). Extrapolating these potential trends until 2050, the viability of grassland use declines and the attractiveness of wetland bioenergy crops increases substantially. At energy prices not much higher than those prevailing in early 2008, assuming modest progress in bioenergy technology, and following a climate scenario where not only higher temperatures prevail but where air circulation patterns have also been affected, reed will outcompete grassland in large areas of the Netherlands. The largest contiguous areas are presently peat land or are covered by partially organic soils.

With our study, we intended merely to explore a method for investigating these questions. The method has some limitations that could be addressed by further research. For instance, our method of assessing viability is based on value added per unit area. This tells us much about the social benefit of a certain land use (whether it would be a worthwhile way of using the land), but less about its commercial feasibility (whether it will pay for the land owner to indeed use the land in that way). Gross value added can be defined as the total value of the produce of the land minus the consumable materials that have been used in the production. To arrive at the net profit per unit area, we need to know the total production cost. In all likelihood, production costs for reed are likely to be low compared to milk (there will be mainly a cost for establishing the reed), but this still needs to be investigated. Our calculations of value added are net of transport and processing, since these are costs beyond the farm gate.

Another aspect that has not yet been included in our study is economies of scale. Transportation costs of reed are high, which means that, in order to operate efficiently, a fairly large production area is needed and the power station being supplied needs to be close by. Smaller patches of land shown on our maps will not, in effect, be feasible production areas.

For the near future, wetland bioenergy crops such as reed and willow are likely to be no more than niche products. In the longer term, however, depending on climate change, energy prices, developments in the dairy market, water management policies in the Netherlands and progress in bioenergy technologies, the transformation of familiar meadow landscapes into wetlands producing reed and willow for electricity generation is far from unlikely.

Acknowledgement We are grateful to the Dutch National Research Programme on Climate Change and Spatial Planning for financing part of the research described here. Another part was funded under the Knowledge Base programme of the Ministry of Agriculture, Nature and

Food Quality. We are, furthermore, grateful to the Netherlands Environmental Assessment Agency (MNP) and the Royal Netherlands Meteorological Institute for providing the modelling framework of Land Use Scanner and parts of the scenario definitions.

References

Banse, M., Van Meijl, H., Tabeau, A., & Woltjer, G. (2008). Will EU biofuel policies affect global agricultural markets? *European Review of Agricultural Economics, 35*, 117–141.

Bertholdsson, N.-O. (2001). Phytoremediation of heavy metals with Salix. In Swedish, English summary. *Journal of the Swedish Seed Association, 111*(2), 84–90.

CPB, MNP, & RPB (2006). *Welvaart en Leefomgeving. Een scenariostudie voor Nederland in 2040* (Prosperity and Living Environment). Centraal Planbureau, Milieu- en Natuurplanbureau en Ruimtelijk Planbureau, The Hague.

Duke, J. (1983). *Handbook of energy crops: Phragmites australis, Ecology*. Published online: http://www.hort.purdue.edu/newcrop/duke_energy/Phragmites_australis.html. Retrieved June 2009.

Greger, M., & Landberg, T. (1999). Use of willow in phytoextraction. *International Journal of Phytoremediation, 1*, 115–123.

Hellings, S. E., & Gallagher, J. L. (1992). The effects of salinity and flooding on phragmites australis. *Journal of Applied Ecology, 29*, 41–49.

Hilferink, M., & Rietveld, P. (1999). Land use scanner: An integrated GIS based model for long term projections of land use in urban and rural areas. *Journal of Geographical Systems, 1*(2), 155–177.

Kok, T. (2004). *Waterberging en natuur. Quick scan naar de combinatie waterberging en natuur (Water storage and nature)*. Ede: Expertisecentrum LNV.

Koomen, E., Loonen, W., & Hilferink, M. (2008). Climate-change adaptations in land-use planning; a scenario-based approach. In L. Bernard, A. Friis-Christensen, & H. Pundt (Eds.), *The European information society; Taking geoinformation science one step further* (pp. 261–282). Berlin: Springer.

Koomen, E., & Van der Hoeven, N. (2008). The Netherlands climate proof; What will the country look like in 2040? *GeoInformatics, 11*(5), 26–27.

Larsson, S., Cuingnet, C., Clause, P., Jacobsson, I., Aronsson, P., Perttu, K., et al. (2003). *Short-rotation Willow biomass plantations irrigated and fertilised with wastewaters. Results from a 4-year multidisciplinary field project in Sweden, France, Northern Ireland and Greece*. Danish Environmental Protection Agency, Sustainable Urban Renewal and Wastewater Treatment Project, Report No. 37, Copenhagen.

LEI (2008). *Land- en tuinbouwcijfers 2008* (Agriculture and horticulture in figures), LEI-report no. 2008-048, LEI, The Hague.

Martin, P. J., & Stephens, W. (2006). Willow growth in response to nutrients and moisture on a clay landfill cap soil. I. Growth and biomass production; II: Water use. *Bioresource Technology, 97*, 437–458.

Riedijk, A., Van Wilgenburg, R., Koomen, E., & Borsboom-van Beurden, J. (2007). *Integrated scenarios of socio-economic and climate change; a framework for the 'Climate changes Spatial Planning' program*, Spinlab Research Memorandum SL-06, Vrije Universiteit Amsterdam.

Rosenqvist, H., & Dawson, M. (2005). Economics of willow growing in Northern Ireland. *Biomass and Bioenergy, 28*(2005), 7–14.

Tielrooij, F. (Ed.). (2000). Waterbeleid voor de 21e eeuw (Water policy for the 21st century), Commissie Waterbeheer 21e Eeuw, The Hague.

Van den Akker, J. (2005). Maaivelddaling en verdwijnende veengronden (Soil subsidence and disappearing peat soils). In W. A. Rienks & A. L. Gerritsen (Eds.), *Veenweide 25x belicht. Een bloemlezing van het onderzoek van Wageningen* (pp. 11–13). Wageningen: Wageningen University and Research Centre.

Van den Hurk, B., Klein Tank, A., Lenderink, G., van Oldenborgh, G. J., Katsman, C., Van den Brink, H., et al. (2006). *KNMI Climate Change Scenarios 2006 for the Netherlands*. KNMI Scientific Report WR 2006-01. De Bilt: KNMI.

Verburg, R., & Jongeneel, R. (August 2008). *Exploring multifunctional land uses as an adaptation strategy to climate change in the Netherlands: An economic assessment of costs and benefits of ecosystem services*. Paper presented to the meeting of the European Association of Agricultural Economists, Ghent.

VROM (2006). *PKB Ruimte voor de Rivier (Room for the river, brochure)*. The Hague: Netherlands Ministry of Housing, Spatial Planning and the Environment (VROM).

WUR-Alterra. (2006). *Dataset Grondsoortenkaart van Nederland 2006* (Soil map of the Netherlands), Wageningen. http://www.bodemdata.nl/. Retrieved 12 August 2009.

Chapter 7
Simulation of Future Land Use for Developing a Regional Spatial Strategy

The Case of the Province of Overijssel

Arjen Koekoek, Eric Koomen, Willem Loonen, and Egbert Dijk

7.1 Introduction

Many geo-information tools, such as visioning, storytelling, forecasting, analysis, sketching, and evaluation, appear to be rarely used for spatial planning (Vonk, Geertman & Schot, 2005). Progress in the application of such tools beyond basic activities, such as researching spatial queries and generating thematic maps, to help solve key spatial planning problems remains limited (Stillwell, Geertman & Openshaw, 1999; Vonk et al., 2005). Land Use Scanner's model has the ability to assist in many of these planning-specific tasks. Especially in scenario-based national forecasts of on the future the model, it has proven to be an adequate tool for informing policy-makers on potential future developments (Borsboom-van Beurden et al., 2005) and has provided ex-ante evaluations of policy alternatives (MNP, 2001; Stillwell et al., 1999; Van der Hoeven, Aerts, Van der Klis & Koomen, 2009). More recently, the model has also been used to optimise projected spatial developments according to specific policy objectives (Borsboom-van Beurden, Bakema & Tijbosch, 2007; MNP, 2007). The resulting maps have the potential to inform policy-makers about alternative solutions for current spatial problems (Koomen, 2008). This chapter demonstrates the capacity of Land Use Scanner to generate optimised spatial developments in actual regional planning contexts. Contributing to the policy formulation process by simulating land use on a regional scale in close cooperation with a regional authority is an interesting step forward in the application of land-use models. Thus, in this case Land Use Scanner has traversed the full spectrum, from academic research to actual planning practice. Applying Land Use Scanner on a regional scale has been a tempting idea since

---

A revised version of this chapter is published as: Koomen, E., Koekoek, A. and Dijk, E. (2011) Simulating land-use change in a regional planning context. *Applied Spatial Analysis and Planning* doi: 10.1007/s12061-010-9053-5. This article is published with open access at Springerlink.com.

A. Koekoek (✉)
Geodan, President Kennedylaan 1, 1079 MB Amsterdam, The Netherlands
e-mail: arjen.koekoek@geodan.nl

E. Koomen, J. Borsboom-van Beurden (eds.), *Land-Use Modelling in Planning Practice*, The GeoJournal Library 101, DOI 10.1007/978-94-007-1822-7_7,

the modelling resolution was increased from 500 to 100 m. Before the currently discussed Overijssel study was done, such applications were, however, limited to a single example (Bouwman, Kuiper & Tijbosch, 2006).

What spatial transformations can be expected in the future and what role can policy-making play in directing developments? The introduction of a new national spatial planning law (*nWRO*) in 2008 created opportunities for the Province to play a more pro-active role in the policy-making arena. In order to visualise spatially accurate versions of possible future scenarios it is paramount to have access to data on the desired spatial level. Scenario simulations rely heavily on the data being input and expert judgement. As many future developments entail a high degree of uncertainty, especially in the longer term, one should always interpret simulations as *possible* future configurations of land use.

7.1.1 A New Regional Spatial Strategy for the Province of Overijssel

In 2008, the strategic plan for the physical environment of the Province of Overijssel that was current at that time needed readjustment, which led to a need for a new strategic plan. Another factor that increased this need was the emergence of new themes on the policy agenda, such as climate change and new insights into demographic trends. The Province of Overijssel's planning authorities – hereafter referred to as the Province – used its new strategic plan to introduce a new planning creed, 'Quality-based decision-making'. The introduction of the new national spatial planning act (*nWRO*) provided a logical moment for redefining the Province's aims and embedding these in spatial plans. For this reason, the Provincial Council decided to develop a new, integral strategy for the physical environment. This plan, called the Regional Spatial Strategy (Omgevingsvisie Overijssel), replaced five plans then current and will legally function as the spatial structural vision, regional water management plan, environmental policy plan, provincial mobility plan and soil-management plan. The Regional Spatial Strategy goes beyond just formulating policy by presenting and framing policy choices and their implementation in an *integral* way for the entire spatial spectrum. The Strategy's sustainability and spatial quality goals function as overarching perspectives that connect the spatial themes it contains.

The Province asked Geodan Next at the start of the policy formulation process to explore new spatial policy alternatives. In this process, important questions were posed: to what degree are current spatial developments 'climate proof', and what are the likely effects of possible spatial policies? Land Use Scanner played an important role in the process of answering questions like these by modelling policy-specific, future land-use configurations at different stages in the development process. From the start of the project, the specific assumptions associated with this type of research were made known to the Province. As many future developments contain a high degree of uncertainty, especially on the longer term, one should always interpret simulations as *possible* future land-use configurations. The spatial images generated can not be used as blueprints of the future: rather, they are to be used to stimulate thinking about cause–effect relationships in policy discussions.

7.1.2 Chapter Contents

This chapter presents two related studies that Geodan Next performed for and in cooperation with the Province. Section 7.2 discusses an exploratory study that provided potential future land-use configurations according to specific policy objectives. The results of this study were used in the policy formulation process for the Regional Spatial Strategy. Section 7.3 describes the Environmental Impact Assessment that Geodan Next subsequently performed for the Regional Spatial Strategy. Finally, Section 7.4 summarises the role of Land Use Scanner in the two studies and finishes with some comments on its general applicability.

7.2 Spatial Exploration: Overijssel in 2040

7.2.1 Creating a Regional Application

The starting point for the regional spatial exploration was the national Second Sustainability Outlook, study, carried out by the Dutch Environmental Assessment Agency (MNP, 2007; see Chapter 4 by Rienk Kuiper, Marianne Kuijpers-Linde, and Arno Bouwman, this volume). In this study, two trend-based scenarios from the national the Welfare, Prosperity and Quality of the living environment study of the Dutch assessment agencies (CPB, MNP & RPB, 2006) were applied to accommodate for uncertainties in demographic and economic growth: one reflecting moderate spatial pressure, comparable to the Transatlantic Market-scenario; and the other reflecting high spatial pressure, similar to the Global Economy scenario. The related scenario assumptions were only used to assess the potential diversity in demand for residential and commercial land. The demand for the other types of land and the spatial relations embedded in the suitability maps are the same for both scenarios, reflecting their trend-based or business-as-usual character. An important assumption in both scenarios in relation to the Province of Overijssel is that the influx of people from the Randstad is limited compared to directly adjacent provinces such as Gelderland, Noord-Brabant and Flevoland.

To apply the national Land Use Scanner application related to the Second Sustainability Outlook to the specific context of Overijssel, several adjustments were necessary. In close cooperation with various representatives of the Province, local suitability variables were adjusted to match current regional spatial policies more closely. The most important adjustments were:

- The importance of national concentration areas (*bundelingsgebieden*) for residential and commercial development was decreased, as this national policy is not considered to be of great importance by the Province.

- Areas close to the larger cities were considered likely locations for residential and commercial development. This applied to a selection of eight relatively large towns in Overijssel. This measure was implemented to limit the growth of smaller towns that was observed in initial national simulations; such growth was considered not to be in line with current regional policies that aim to concentrate urbanisation.
- Commercial areas from the IBIS database were considered attractive for new commercial development.
- New developments in flood-prone areas and water-retention areas were discouraged through restrictive policies.
- A spatial redefinition of the ecological main structure (the national plan for a spatially coherent set of nature areas) to replace an older version. New commercial and residential functions in the ecological structure and national landscapes were considered unlikely and received a higher negative suitability factor.

The above adjustments were incorporated in the suitability maps that describe the preferred locations for the various types of land use. Restrictions lead to lower or even negative land-use suitability values, whereas stimulating policies are reflected as higher suitability values. Application of the adjusted suitability values resulted in adjusted spatial patterns for Overijssel in the year 2040.

The Province was particularly interested in the spatial impact of the trend-based scenario simulations on certain sensitive areas. By using an interactive set-up in a workshop, the Province was able to provide feedback on preliminary results and thus fine-tune the resulting land-use patterns. The impact of urbanisation was analysed in relation to three different policy themes. The rationale behind these impact assessments is the following: from a nature perspective, building in the ecological main structure is discouraged; for landscape conservation, from a landscape preservation perspective large-scale urban development in the national landscapes is discouraged; and from a flood-risk perspective, building in flood-prone areas is unwanted. The impacts were assessed by comparing pixel-by-pixel the urban (residential, commercial and greenhouses) locations shown in the maps of current and simulated land use. The new urban locations were then overlaid on maps representing the specific policy themes. This procedure showed that new urban development in nature areas and national landscapes is limited in the two trend-based scenarios, although such development is more abundant within flood-prone areas, especially near the large cities of Zwolle and Deventer. Especially in the scenario reflecting high spatial pressure, the amount of urban development in the flood-prone areas is substantial.

7.2.2 Visualising Theme-Specific Spatial Policy Alternatives

What spatial developments can be expected if one thematic ambition is taken as the leading objective in spatial planning? By giving one policy objective overriding importance in land-use simulations, the spatial impact of successful implementation

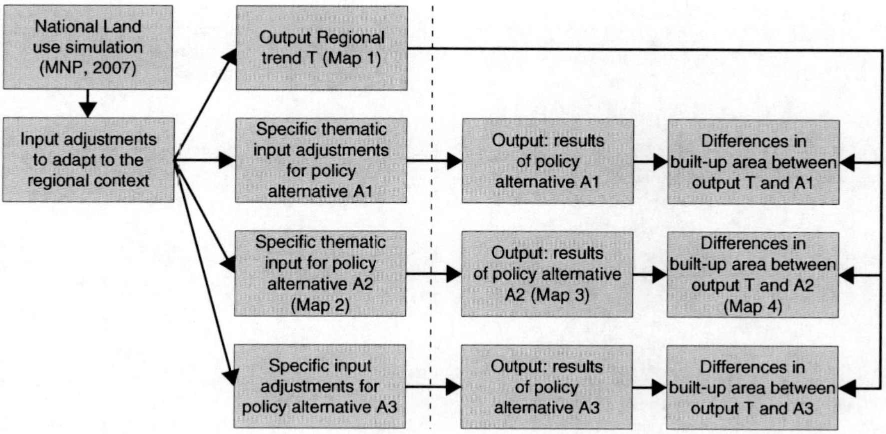

Fig. 7.1 Flowchart describing the creation of regionalised policy-specific land-use simulations. Maps 1–4 are included in Fig. 7.2

of that specific policy can be shown. The Province was interested in the potential spatial developments related to three different spatial policy ambitions. These policy alternatives were formulated as coherent sets of spatial policies related to the themes of: (1) water management; (2) safety and health; and (3) compact urbanisation. The high spatial pressure scenario was used for these simulations, because it maps the impact of the policy measures more clearly. As such, it prepares policy-makers for a worst-case scenario with respect to the spatial developments that had to be accommodated in the policy alternatives. The regional model application described in the preceding section was used as a starting point in this optimisation effort. For the simulations of the three thematic policies, additional, policy-related spatial datasets were added in Land Use Scanner to stimulate or restrict certain developments. By comparing the simulation results of the trend-based scenario (T) and the theme-specific policy alternatives ($A_{1,2,3}$), spatial differences could be visualised, as described by Fig. 7.1.

The first thematic policy alternative focused on the optimisation of spatial developments from a water management perspective. Several policies are included in this alternative. To prevent drought, groundwater protection areas are kept free from urban development. To limit the potential impact of flooding, urban development is not allowed within flood-prone areas, near major waterways or in areas with a high risk of inundation, such as winter riverbeds. Compared with the trend-based scenario, this implies a shift in development from the lower-lying cities of Zwolle and Kampen towards cities and villages situated on slightly higher ground, such as Steenwijk and Hardenberg.

A second thematic policy alternative puts health and safety policies first in new spatial developments. Residential development is therefore restricted in areas with high levels of noise pollution, high levels of personal risk (e.g. near dangerous plants, transportation routes for dangerous goods, gas lines or high-voltage power lines). Figure 7.2 shows the development of this specific policy alternative and

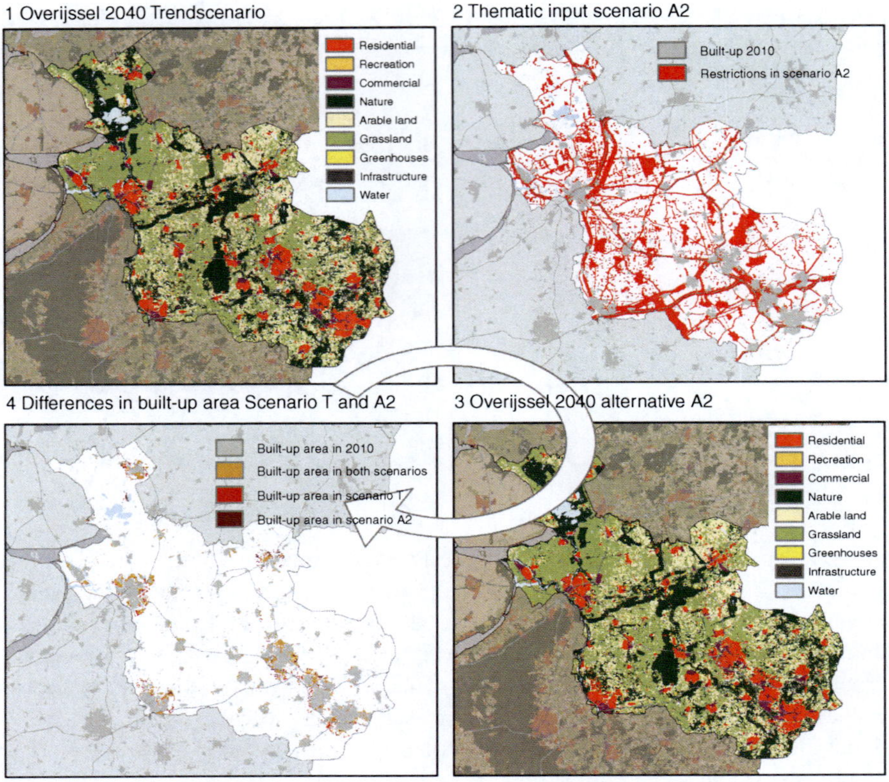

Fig. 7.2 The four main steps (clockwise from *top left*) for analysing the spatial impact of the health and safety policy alternative

compares its distribution of urban areas with that of the trend-based scenario. Crucial in this alternative are the areas with safety and health restrictions that are safeguarded from new developments by adding negative suitability factors. The resulting maps show developments near major roads coming to a halt, in turn leading to increased urban development near some smaller cities. Above all, however, the analysis shows that enough alternative urbanisation locations are available to implement this extensive set of environmental policies.

A third thematic policy alternative focused on compact urbanisation strategies. One such strategy could be to spread residential development among more and smaller cities. Another could be to centralise residential developments in the largest cities.

In short, this study revealed that the projected urbanisation could result in local conflicts regarding landscape quality, personal safety and water management. The Province used the spatial exploration to frame the possible effects of new building developments on the ecological main structure, the preservation of the national landscapes and on flooding risks. The study provided with insights that the Province used in the process to formulate policies for its Regional Spatial Strategy.

7.3 Strategic Environmental Assessment

The Province chose to develop a Regional Spatial Strategy rather than produce separate plans for spatial development, water management, mobility, soil and environment. Because this strategy sets conditions for other plans and activities related to, for example infrastructure, rural developments and Natura 2000 areas, the law requires a Strategic Environmental Assessment (SEA) to be made.

7.3.1 Sustainability Indicators

The Province decided to expand the scope of this SEA from a strictly environmental impact assessment (EIA), to that of a more encompassing sustainability impact assessment. Sustainable development is perceived here as a balanced development of human, natural and economic capital. These three aspects are often referred to as the 'People', 'Planet' and 'Profit' dimensions of sustainability (Elkington, 1994; Hermans & Dagevos, 2006). The policy impacts for these sustainability dimensions were assessed in the SEA for the proposed Regional Spatial Strategy in comparison with the current policy. We discuss the applied indicators in more detail in Section 7.3.3.

At the start of the assessment process, the EIA commission advised to: develop a list of indicators with which to assess the sustainability impacts; perform analyses to reveal current policy shortcomings and dilemmas; elaborate on the role of the Province; and specify impacts for Natura 2000 areas. To do so, a list of indicators – qualitative and quantitative – for the People, Planet and Profit aspects of sustainability was formulated; whenever possible, the use of quantitative, reproducible methods was preferred above qualitative methods. Apart from assessing sustainability impacts, the SEA also entailed providing input for the policy formulation process. This encompassed, amongst other things, an analysis of the current policy shortcomings. During the assessment process, workshops were organised to present preliminary results and to introduce possible policy alternatives.

The SEA study was conducted simultaneously with the formulation of the Regional Spatial Strategy. This had the advantage that preliminary assessment results in the SEA could be used by the Province as input during the process of formulating policy for the Regional Spatial Strategy. This iterative working process came at the price of requiring regular updates of the impacts on sustainability.

7.3.2 Developing the Current and New Policy Alternatives

Land Use Scanner was used in the SEA process to simulate future land use according to the conditions under current policy and those under policies outlined in the new integrated regional plan (i.e. the Regional Spatial Strategy). These simulations were the main input for the assessment of spatially-explicit sustainability impacts. The assessment of non-spatial impacts is not discussed in this chapter.

Two land-use simulations were carried out: one based on the continuation of current policy (Current Policy alternative), and one based on the introduction of the Regional Spatial Strategy (New Policy alternative). The regionalised national simulations we describe in Section 7.2.1 were the starting point for these simulations. To simplify the assessment process, only one scenario was applied: the moderate spatial pressure scenario was selected as the demographic and economic developments underlying this scenario are considered to be more probable than those associated with the high spatial pressure scenario. A slightly revised version of the trend-based scenario of the regional exploration study was used as the Current Policy alternative. This revision mainly concerned a small decrease in the demand for residential and commercial land in line with the most provincial policies (Koomen et al., 2008b). In the New Policy Alternative several additional stimulations and restrictions were added to the Current Policy Alternative, partly replacing initial policies:

- a spatial redefinition of the ecological main structure was incorporated that includes more detailed spatial demarcation and some areas, important from a water quality perspective;
- the current locations of greenhouses were maintained, although locally some expansion is possible;
- new developments in commercial transportation were stimulated on sites that are well-connected to international water, rail and road networks;
- the compact urbanisation strategy introduced in the spatial exploration project and discussed in preceding section was chosen in the New Policy alternative, thus concentrating residential developments in the larger towns;
- recreational developments were stimulated in attractive rural areas, nearby the ecological main structure or the national landscapes.

The differences between the Current and New Policy alternatives are considerable, especially for the location of new urban areas (highlighted in Fig. 7.3).

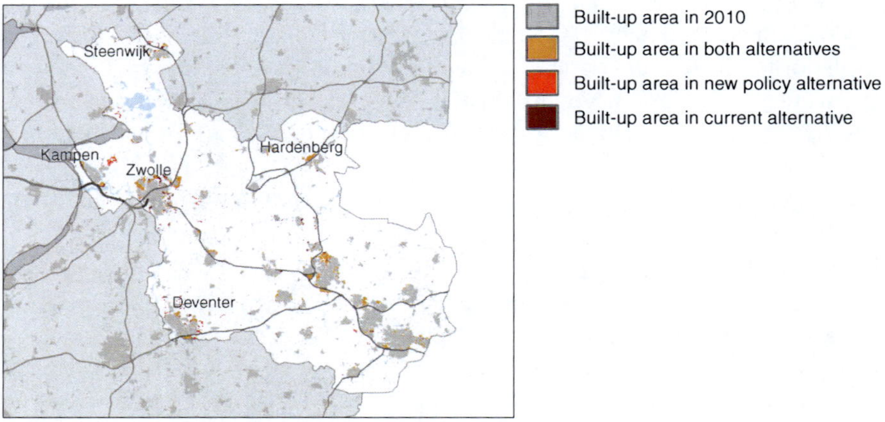

Fig. 7.3 Increase in built-up area by 2040 under current and new policy alternatives

7.3.3 Assessing Environmental Impacts

To what degree are policy aims regarding sustainability met in the Current and New Policy alternatives? In the SEA, a list totalling 28 indicators for the domains of People, Planet and Profit was used to assess whether the current policies are sufficient to meet the sustainability aims. Environmental impacts were calculated using several approaches, depending on the availability of established spatial data and methods of analysis that could be applied within the time frame of the SEA process. Table 7.1 presents a selection of 12 indicators that, to some extent, relied on land-use information from Land Use Scanner. These indicators link to policy themes for which the spatial distribution of land use is important.

In most SEAs, impacts are described in a qualitative manner based on expert judgement (see, for example, VROM, 2008). This makes the assessment process more difficult to reproduce, less transparent and potentially sensitive to subjectivity. In the SEA of Overijssel's regional strategy we limited these problems as much as possible, striving to maximise reproducibility, transparency and objectivity in the impact assessment. Land Use Scanner played a crucial role in this respect, providing a quantitative basis for the assessment of many sustainability impacts. The degree to which the model could be used in the assessment process is indicated in Fig. 7.4.

When these were available, we preferred using Land Use Scanner's internal modules for the impact assessment. As these internal modules were designed to utilise simulation outputs, no additional data transformations were necessary. However, for most sustainability indicators no internal modules in Land Use Scanner are available as yet. In these cases, the land-use configurations can often be used as inputs for an external assessment model or for a quantitative comparison.

Table 7.1 Overview of sustainability indicators related to land-use information from Land Use Scanner

Sustainability theme/Aspect/Indicator	Relation to land use scanner
People/Landscape/Landscape openness	Internal land use scanner assessment model
People/Landscape/Preservation national landscapes	Quantitative comparison
People/Landscape/Identity and diversity	Quantitative comparison
People/Safety and health/Flood risk	Internal land use scanner assessment model
People/Safety and health/Excess water damage	Quantitative comparison
Planet/Nature/Condition Natura 2000 areas	Input for external assessment model
Planet/Nature/Realisation EHS	Quantitative comparison
Profit/Economy/Zero-grazing livestock area	Quantitative comparison
Profit/Accessibility/Car accessibility urban areas	Input for external assessment model
Profit/Accessibility/Public transport accessibility urban areas	Input for external assessment model
Profit/Accessibility/Private motor vehicle accessibility rural areas	Input for external assessment model
Profit/Accessibility/Public transport accessibility rural areas	Input for external assessment model

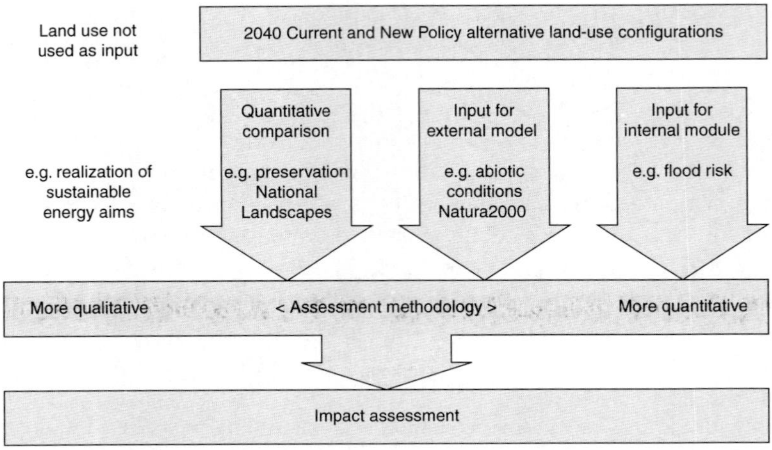

Fig. 7.4 Different applications of Land Use Scanner output for impact assessment in the SEA

To illustrate the process of assessing sustainability indicators in the SEA, we provide below an example of each of the three types of impact assessment that were performed in relation to the land-use simulation results.

An example of an *internal assessment module* that is available in Land Use Scanner is the analysis of flood risk. This module was developed through the cooperative efforts of a number of parties, including the Environmental Assessment Agency, Delft Hydraulics and the VU University Amsterdam, and made available for this project through the Climate *changes* Spatial Planning Programme (www.klimaatvoorruimte.nl). Future land-use configurations were used in this module to assess the potential economic damage and number of casualties that may result from flooding (see Van der Hoeven et al., 2009). The results of the assessment reveal that the differences between the two policy alternatives are negligible. A more complete overview of all the land-use-related indicators that can be calculated within the Land Use Scanner model is available in the literature (Bubeck & Koomen, 2008).

When no internal assessment model is available, land-use patterns can be input in *external analysis tools or models*. This is, for example, the case for the analysis of coherence and abiotic conditions in Natura 2000 areas; achieving favourable conditions in nature areas is an important policy goal. Geodan Next developed a systematic approach to assess the impact of various spatial developments on basic conditions in Natura 2000 areas. Through a predefined set of spatial operations, made operational in an ArcGIS script, several aspects of Natura 2000 areas were analysed. These relate to the coherence of existing and newly created nature areas; differences in nitrogen deposits in these areas; drought effects; and adverse building and other developments in those areas that have a hydrological impact on Natura 2000 areas. The simulations of land use provided information on the expected future configuration of nature and new urban areas that was used as one of the input variables for this assessment approach.

If land-use changes are deemed relevant for a sustainability policy issue, but internal and external assessment models are lacking, Land Use Scanner output can be used for *quantitative comparison.* This is, for example, the case for the indicator 'preservation of national landscapes'. National and regional policies aim to preserve the 'core qualities' of these landscapes by, amongst other things, limiting large-scale urbanisation. Therefore, new urban areas in the two national landscapes of Overijssel were isolated from the land-use simulations and compared for both the trend and policy alternatives. This spatial analysis highlighted differences in the amount and location of projected urban development. The New Policy alternative contained slightly more building development in the national landscapes than the Current Policy. Nevertheless, this difference was considered too marginal to be included in the final assessment.

In cases for which land-use changes are not important for achieving policy objectives, simulation output was not used in the assessment process. This is the case for issues such as the realisation of sustainable energy aims.

7.4 Discussion and Conclusion

Land Use Scanner proved to be a useful tool for supporting the regional planning process in the Dutch Province of Overijssel completed in 2009. Different capacities of Land Use Scanner were used in two types of study. The spatial exploration study focused on the optimisation of land-use patterns according to rather hypothetical sets of policy objectives, whereas the Environmental Impact Assessment relied on the model's ability to provide probable land-use patterns related to an integrated policy alternative that could be used as input in various impact assessments. Both studies contributed to the acceptance and formulation of a Regional Spatial Strategy for the province. This section briefly discusses experiences with using the model in the two studies and concludes with some comments on its general applicability in spatial planning processes.

In the spatial exploration study, Land Use Scanner was used to frame current spatial policy objectives and simulate land-use patterns according to coherent sets of policy objectives. In cooperation with the Province, regional policy maps were used to improve the quality of data input and create regional future land-use configurations linked to two trend-based scenarios. These offered a solid framework for developing and comparing the sets of spatial policy alternatives. As the same economic, demographic and other scenario assumptions – and thus the same quantities of land-use change – were used in both simulations, they offer a clear view of the specific implications of the spatial policy alternatives. Indeed, comparison revealed potential (unwanted) spatial impacts directly related to the proposed policy sets. The use of two trend-based reference scenarios allowed the assessment of these impacts under different levels of spatial pressure. The inclusion of more diverse reference scenarios would have provided an even more robust assessment of the impacts of the sets of policy alternatives. But this benefit would have come at the cost of obscuring the specific impacts associated with these alternatives since more variation in the spatial developments would be introduced. An underestimated, but significant, benefit of the modelling process lies in its iterative, open character. The application of Land Use Scanner and the presentation of intermediate results in workshop sessions provided a forum that encouraged people from a variety of professional backgrounds (e.g. water and nature management, urban development) to discuss potential developments, propose alternative policies and, above all, clarify their suggested planning interventions in a shared spatial environment. In the process of constructing a coherent Regional Spatial Strategy, Land Use Scanner provided a means for exploring and integrating sector-specific interests and thus aided the policy formulation process.

The SEA of the Regional Spatial Strategy for Overijssel is unique since it is the first in the Netherlands in which land-use modelling results were used in the sustainability assessment process. SEAs are often fairly qualitative and thus tend to be non-transparent in nature. The use of Land Use Scanner made part of the sustainability assessment process more transparent and reproducible. The results of the SEA and the role Land Use Scanner played in the assessment process were greatly appreciated by the provincial policy-makers involved. They judged the sustainability impacts to be presented in a clear manner, despite the complex assessment process. It is interesting to note that the policy-makers preferred to use only one trend-based reference scenario as they found the addition of an extra scenario to be confusing.

The studies also revealed two more general modelling issues that have also been encountered in subsequent studies performed for other provinces (Atzema et al., 2008; Kuijpers-Linde et al., 2008). These relate to the impact that can be expected of planning initiatives and the need for integrated data infrastructures. Each of these issues is briefly explained below.

On the issue of impact, a prominent feature of the land-use simulations we present in this chapter is the assumption that the proposed spatial policies will be actually applied and fully effective. Land Use Scanner thus shows what could happen if a certain policy is put in place. In practice, restrictive or stimulative spatial policies are not often as effective as intended. Even spatial policies that are generally considered successful, such as the national buffer zones intended to keep green zones between major cities free from urban development, do not fully prevent urbanisation (Koomen et al., 2008a). This issue of effectiveness is especially relevant when the aim of the study is to assess the impact of a specific policy. The model has the flexibility to account for partial effectiveness of implemented policies, but quantitative assessments of the degree of success of different types of spatial policies are rarely found in planning literature. This indicates a more generally felt lack of ex-post evaluations of spatial plans and concepts. The dynamic, interrelated and often vague nature of most spatial policies and plans makes this a difficult issue, of course, and establishing the effectiveness of new planning concepts that have no historic parallel is obviously close to being impossible. Yet, more attempts in this direction would greatly enhance planning in general and our type of land-use modelling in particular.

Furthermore, consistent and up-to-date data input are a prerequisite for successful modelling. In our study the data needed were scattered across different governmental departments and stored in range of different data formats. Some departments would even use different socio-economic scenarios for various future sector-specific outlooks. Comparable studies performed for other provinces have provided similar experiences. We hope that the emphasis regional authorities currently put on integrated (spatial) data-infrastructures will improve data consistency and availability, thus enhancing the potential for spatial analysis in general and land-use modelling in particular.

To conclude, we propose that Land Use Scanner be developed in such a way that it becomes a platform for integrated spatial-impact assessment that makes sustainability assessment more transparent and robust. For many sustainability indicators with a spatial component, model output still plays a modest role, while we observe a need for a more quantitative assessment base: only a couple of internal Land Use Scanner assessment modules were available for the SEA conducted. Development of new modules for, for example, accessibility, safety and health (e.g. air pollution and external safety) would significantly increase the value of Land Use Scanner as a platform for sustainability assessment.

References

Atzema, O., Van Egmond, K., Mommaas, H., Wenting, R., & Kuijpers-Linde, M. (2008). *Utrecht 2040; Strategische notities in het kader van het traject 'Samen op Weg naar 2040' van de Provincie Utrecht*. Utrecht: Universiteit Utrecht Faculteit Geowetenschappen.

Borsboom-van Beurden, J. A. M., Bakema, A., & Tijbosch, H. (2007). A land-use modelling system for environmental impact assessment; Recent applications of the LUMOS toolbox. Chapter 16. In E. Koomen, J. Stillwell, A. Bakema, & H. J. Scholten (Eds.), *Modelling land-use change; Progress and applications* (pp. 281–296). Dordrecht: Springer.

Borsboom-van Beurden, J. A. M., Boersma, W. T., Bouwman, A. A., Crommentuijn, L. E. M., Dekkers, J. E. C., & Koomen, E. (2005). *Ruimtelijke Beelden; Visualisatie van een veranderd Nederland in 2030*. RIVM report 550016003. Bilthoven: Milieu- en Natuurplanbureau.

Bouwman, A.A., Kuiper, R., & Tijbosch, H. (2006). *Ruimtelijke beelden voor Zuid-Holland*. Rapportnummer 500074002.2006. Bilthoven: Milieu- en Natuurplanbureau.

Bubeck, P., & Koomen, E. (2008). *The use of quantitative evaluation measures in land-use change projections; An inventory of indicators available in the land use scanner. Spinlab Research Memorandum SL-07*. Amsterdam: Vrije Universiteit Amsterdam/SPINlab.

CPB, MNP and RPB (2006). *Welvaart en Leefomgeving*. Een scenariostudie voor Nederland in 2040. Den Haag: Centraal Planbureau, Milieu- en Natuurplanbureau en Ruimtelijk Planbureau.

Elkington, J. (1994). Towards the sustainable corporation: Win-win-win business strategies for sustainable development. *California Management Review, 36*(2), 90–100.

Hermans, F., & Dagevos, J. (2006). *De duurzaamheidsbalans van Brabant 2006*. Tilburg: Telos.

Koomen, E. (2008). *Spatial analysis in support of physical planning*. VU University: Amsterdam.

Koomen, E., Dekkers, J., & Van Dijk, T. (2008a). Open space preservation in the Netherlands: Planning, practice and prospects. *Land Use Policy, 25*(3), 361–377.

Koomen, E., Loonen, W., & Koekoek, A. (2008b). *Beschrijving en uitwisseling regionale Ruimtescanner toepassingen*. Amsterdam: Geodan Next.

Kuijpers-Linde, M., Koekoek, A., & Loonen, W. (2008). *Uitwerking ruimtelijke beelden voor het nieuwe omgevingsbeleid van Drenthe*. Amsterdam: Werknotitie. Geodan Next.

MNP (2001). *Who is afraid of red, green and blue? Toets van de Vijfde Nota Ruimtelijke Ordening op ecologische effec*ten. RIVM-rapport 711931005. Bilthoven: RIVM.
MNP (2007). *Nederland Later; Tweede Duurzaamheidsverkenning deel fysieke leefomgeving Nederland*. MNP-publicatienr.500127001/2007. Bilthoven: Milieu- en Natuurplanbureau.
Stillwell, J. C. H., Geertman, S., & Openshaw, S. (1999). *Geographical information and planning. Advances in spatial science*. Berlin/Heidelberg/New York: Springer.
Van der Hoeven, E., Aerts, J., Van der Klis, H., & Koomen, E. (2009). An integrated discussion support system for new Dutch flood risk management strategies. In S. Geertman & J. C. H. Stillwell (Eds.), *Planning support systems: Best practices and new methods* (pp. 159–174). Berlin: Springer.
Vonk, G., Geertman, S., & Schot, P. (2005). Bottlenecks blocking widespread usage of planning support systems. *Environment and Planning A, 37*, 909–924.
VROM (2008). *PlanMER Structuurvisie Randstad 2040: naar een duurzame en concurrerende Europese topregio*. Den Haag: Ministerie van Volkshuisvesting, Ruimtelijke Ordening en Milieubeheer (VROM).

Chapter 8
Lessons Learned from Using Land-Use Simulation in Regional Planning

Chris Jacobs, Arno Bouwman, Eric Koomen, and Arjen van der Burg

8.1 Introduction

What happens if we maintain current spatial planning policies? How will new planning concepts turn out? Where is potential for urban expansion and where are valued landscapes under threat? These are questions a planner might often ask. Models that simulate future land use, such as Land Use Scanner, may provide answers to such questions. Until recently, such land-use simulations in the Netherlands were mainly performed on the national level, as described in Chapters 4 and 5 of this book. Studies such as these were often done *in preparation for* or *as an evaluation of* national spatial-planning decisions.

The revised Land Use Scanner model, with its more detailed 100-m grid resolution, is now more suitable for regional planning applications. The model's potential to simulate regional developments was first explored by PBL in 2006 in cooperation with the province of Zuid-Holland (see Borsboom-van Beurden, Bakema & Tijbosch, 2007; Bouwman, Kuiper & Tijbosch, 2006). Several other regional studies followed in 2008. This chapter briefly describes three types of regional applications of Land Use Scanner. The first type was performed by Geodan Next in conjunction with PBL and was used to *provide input for regional strategic visions* of three Dutch provinces (see Section 8.2). A second type, performed by the Dutch Ministry of Housing, Spatial Planning and the Environment (VROM), was aimed at *exploring spatial planning concepts* in the preparatory phase of the 'Randstad 2040' project (see Section 8.3). In the third type, the VU University Amsterdam used Land Use Scanner *to integrate sector-specific climate adaptation measures* for the Province of Groningen (see Section 8.4). Although each of the three types of projects had different objectives, all made use of Land Use Scanner and these experiences are discussed in this chapter with reference to:

C. Jacobs (✉)
Department of Spatial Economics/SPINlab, VU University Amsterdam,
De Boelelaan 1105, 1081 HV Amsterdam, The Netherlands
e-mail: c.g.w.jacobs@vu.nl

E. Koomen, J. Borsboom-van Beurden (eds.), *Land-Use Modelling in Planning Practice*, The GeoJournal Library 101, DOI 10.1007/978-94-007-1822-7_8,

- the purpose of the project;
- the kind of spatial developments that were depicted;
- the policy alternatives that were considered; and
- the modelling- and planning- related lessons learned.

8.2 Providing Input for Regional Strategic Visions

In 2008, a new Dutch spatial planning act came into force. In this new act, municipal, provincial and national administrations are obliged to create a strategic vision as a guiding 'master plan' for legally binding spatial plans and other policy measures (e.g. providing subsidies). Furthermore, within the boundaries of national policies that are formalised in a spatial by-law, the act gave regional authorities greater responsibilities in the planning process. The new spatial planning act demands the definition of strategic regional and local visions that are new in legal status and in the magnitude of effect on local policies. In 2008, a number of provinces (Dutch regional administrations) began defining these strategic visions, based on almost a century of regional planning tradition. Geodan Next supported the development of these visions by providing simulations of future land use.

Spatially explicit depictions of future spatial developments in a province, based on current socio-economic trends and spatial policies, offered a useful and thought-provoking starting point for the development of strategic visions. These depictions were based on assumptions and quantified explorations of trends taken from the 'Second Sustainability Outlook on the future of the Netherlands' study of PBL (MNP, 2007), which is also discussed in Chapter 4. In the provincial explorations, national spatial depictions were complemented with province-specific information. Such information included regional policies on restrictions and incentives, more detailed figures on the demand for land per capita and region-specific knowledge on regional spatial developments.

In all studies, simulated autonomous developments were shown on maps of future land-use patterns related to two trend-based scenarios. Both scenarios attach the same importance to current spatial policies, but they differ in the demand for space. The simulations show, therefore, potential autonomous developments under conditions in scenarios of moderate development pressure (*Baseline scenario*) and high development pressure (*High Development Pressure scenario*) previously used in the 'Second Sustainability Outlook' study. Besides the land-use maps, many other maps were produced to depict possible impacts of the projected spatial developments. These maps focus on themes such as urban sprawl, deterioration of landscape values and flood risk. For a more elaborate description of these maps studies, see relevant publications (Atzema, Van Egmond, Mommaas, Wenting & Kuijpers-Linde, 2008; Koomen, Kuijpers-Linde & Loonen, 2008b; Kuijpers-Linde, Koekoek & Loonen, 2008) and Chapter 7 by Arjen Koekoek, Eric Koomen, Willem Loonen, and Egbert Dijk, this volume.

For the provinces of Overijssel and Drenthe, the depictions of autonomous spatial developments were supplemented with studies on the possible spatial consequences

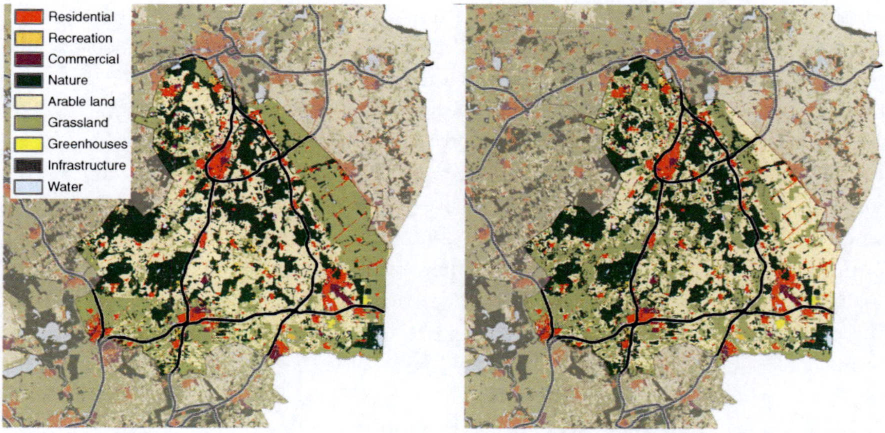

Fig. 8.1 Land-use simulations for the province of Drenthe in 2040 according to employment (*left*) and climate adaptation policy alternatives (Source: Kuijpers-Linde et al., 2008)

of new policy alternatives. Together with this supplementary work, the studies show not only the tasks at hand for these provinces but also explore possible solutions. The simulations of alternative policies for Drenthe and Overijssel were only done for the *High Development Pressure scenario* as this was thought to more clearly depict potential developments: the simulations show a kind of worst-case scenario with regard to environmental impact.

By way of example, Fig. 8.1 shows two policy alternatives that were simulated for the province of Drenthe. The maps show the potential spatial patterns associated with a strong focus on employment (stimulating, for example, a concentration of industrial/commercial activities and competitive large-scale farming) and climate adaptation (e.g. promoting robust nature areas and pastures in small river valleys). (Other policy alternatives stressed the recreational and natural values of the province.) Land-use simulations were also developed for the province of Overijssel that stressed specific policy ambitions. Elements of these, often rather extreme, policy alternatives were eventually included in the final policy alternative of the strategic vision. In the compulsory environmental impact assessment (EIA) of the regional plan, this policy alternative was assessed on sustainability aspects, which are briefly described in Chapter 7.

8.2.1 Lessons Learned

Trend-based simulations were used to understand which issues should be on policy-makers' agendas. In addition, simulations of policy alternatives were developed to explore possible planning alternatives. The simulation results were highly praised by the provincial authorities. According to the policy makers involved, the simulations clearly showed the extent of expected spatial developments, making the simulations useful for estimating potential impacts of spatial policies. But not only the final land-use maps proved to be useful. Especially discussions about intermediate results were highly valued by the provincial policy-makers: these helped them understand and prioritise issues on their spatial planning agenda. For example, the environmental effects of light pollution initially seemed important for the regional planners of Overijssel. However, during discussions about the land-use modelling results, the spatial impacts of light pollution proved hard to define and were, in general, relatively small compared to the impacts of urbanisation and urban sprawl. Subsequently, light pollution lost priority to other issues in the Overijssel modelling process.

The simulations raised a number of research questions and led to suggestions for model improvements that deserve more attention in following projects. One important question relates to the demand for space that is specified in the trend-based scenarios used in the simulations. These figures for demand were initially taken from the 'Second Sustainability Outlook' study (MNP, 2007). However, the figures in that study are for expected demand for residential land at the national level, not at regional level. Furthermore, the evaluation of local suitability values for specific types of land use might be improved if the balance between factors such as accessibility, quality of the neighbourhood and policy restrictions is thoroughly founded on empirical findings. This understanding may help identify more precisely future developments in urbanisation.

Experiences from the provincial studies point to a number of improvements of the modelling tools used. Ideally, a direct link between Land Use Scanner and the models that produce the sector-specific demand for land should be established. By that means, direct feedbacks could be applied between consequences of spatial developments and regional demands for land. This would enhance Land Use Scanner's capacity to match developments between, for example, the housing market and markets for industrial and commercial land.

Currently, Land Use Scanner only simulates mono-functional types of land use. Discussions with policy-makers have made it clear that the combination of different types of use on the same piece of land is an increasingly popular approach for reconciling the many different demands for space. Incorporation of multifunctional land-use types in the model is thus desirable to be able to simulate such novel approaches relating to, for example, nature management on agricultural land or water management in nature areas. Furthermore, in a number of simulated policy alternatives, accessibility measures and related developments were deemed important. To facilitate simulations that focus on the impact of accessibility on transport-related policy alternatives, further integration of Land Use Scanner and a transport model ought to be considered.

8.3 Exploring Spatial Planning Concepts for the Randstad Conurbation

The urban core in the Netherlands, known as the 'Randstad', presents very specific spatial planning challenges. The special status of the Randstad conurbation has been acknowledged at a national level by Dutch policy-makers, resulting in the need for an integrated strategic plan for spatial developments there. Consequently, the Dutch Ministry of Housing, Spatial Planning and the Environment (VROM) launched the so-called 'Randstad 2040' study in 2008. The VU University Amsterdam used Land Use Scanner for an initial depiction of various regional planning concepts, whereas VROM applied Land Use Scanner to perform a more elaborate exploration of the regional possibilities for urban densification and expansion. Both simulations were used as input for the 'Randstad 2040' study.

8.3.1 Depicting Regional Planning Concepts

At the very start of the 'Randstad 2040' study, three spatial concepts for the Randstad conurbation were publicised by the Dutch minister responsible for spatial planning (Cramer, 2008). These concepts were seen as possible solutions for spatial problems and were to serve as outlines for the further spatial development of the Randstad conurbation. The concepts were:

- extending current city boundaries;
- concentrating urban development along major transport corridors; and
- intensifying current land use within existing city boundaries.

The SPINlab of the VU University Amsterdam simulated future land use while taking these planning concepts into account. The simulations were adaptations taken from the 'Second Sustainability Outlook' study (MNP, 2007). SPINlab's goal in doing so was to depict these concepts, taking into account the amount of land needed for urban use in the Randstad conurbation. This enabled policy-makers to understand the extent of demand for urban land in the area and helped create a sense of urgency for finding solutions to the problems that accompany that demand.

Such a sense of urgency can be created when the severity of the situation is shown, so a worst-case scenario was used in simulating the effects of the spatial policy concepts listed above. Many problems in the Randstad stem from lack of space for urban land use and, therefore, the worst-case scenario is one in which demand for urban land is highest. Consequently, the simulations performed assumed the same *High Development Pressure scenario* used in the 'Second Sustainability Outlook' study. In this scenario, large portions of land in the Randstad conurbation are presumed to be needed for new residential, industrial and commercial use.

In the simulations, the impacts of putting the three concepts into practice had to be translated into logic rules and values to be used in Land Use Scanner. This was done in cooperation with an expert from VROM, the ministry involved in the 'Randstad 2040' study. Expert judgement was used instead of academic exploration of these logic rules and values because the planning-process dynamics demanded relatively fast results. A rapid and clear depiction of planning concepts and the regional demand for urban land was deemed more important than a thorough analysis and assessment of all the location factors and spatial processes involved.

8.3.2 Lessons Learned

Use of Land Use Scanner in this initial phase showed that the model is a powerful tool for quickly visualising planning concepts with specific restrictions. Especially depictions in which the planning concept and land-use simulation results were combined (see Fig. 8.2) turned out to be very communicative. It should be noted, however, that the validity of such applications is limited as they are strongly influenced by the relative values the modeller applies to the various components of the suitability values used when producing visualisations of planning concepts. They do not show the likely consequences of planning concepts, but rather visualise planning concepts that would otherwise remain rather vague, qualitative notions. While the land-use simulations were being done, the preference of VROM turned towards one of the policy concepts explored here: intensifying land use within current city boundaries. This brought with it the need to incorporate more detail on the impact of land-use densification in the model. The resulting simulations, which were performed by VROM itself, were used as input in the 'Randstad 2040' study (described in the following subsection).

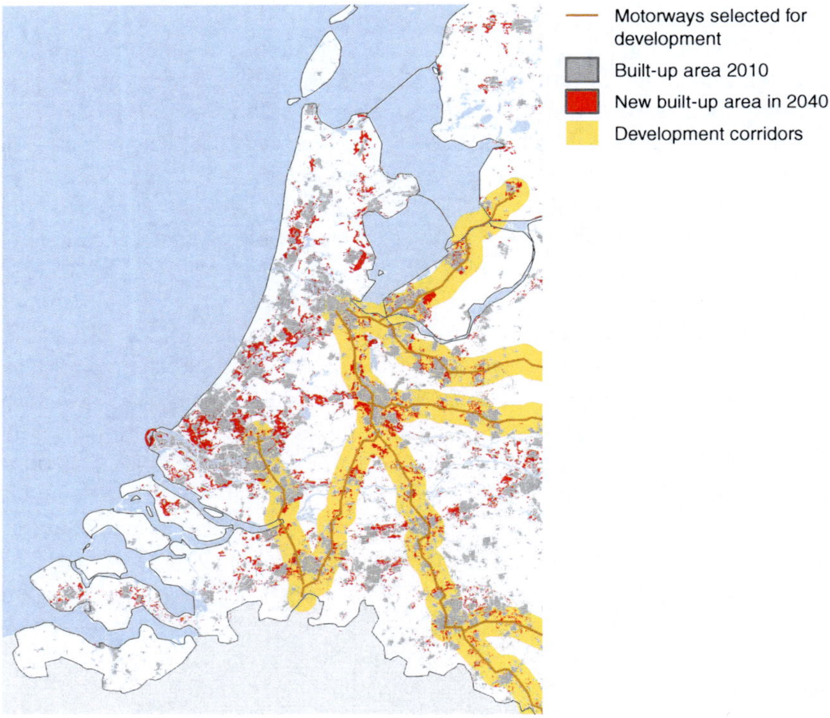

Fig. 8.2 Depiction of new urban areas in the transport axes corridor concept. *Orange* indicates the corridors in which urban development was simulated

8.3.3 Exploring Regional Possibilities for Urban Densification and Expansion

For the development of a structural vision for the 'Randstad 2040' study, the Ministers of Housing, Spatial Planning and the Environment (VROM), Economic Affairs (EZ), Transport, Public Works and Water Management (V & W) and Agriculture, Nature and Food Quality (LNV) agreed upon a number of main assumptions relating to the long-term spatial development and policy priorities of the area:

1. A large permanent demand for land is expected;
2. The quality of the demand for land is becoming increasingly important;
3. Specialisation, and (spatial) differentiation are important in realising economic growth objectives;
4. Internal and external accessibility must be substantially improved;
5. Improvement of public transport is very important to improve accessibility;
6. More space is needed for safety (flooding risks near the coast and rivers) because of the future effects of climate change;
7. Urbanisation should be concentrated, but land is needed to create spacious green living and working environments;
8. A focus on major restructuring and transformation of the existing built-up area.

These priorities are described in the 'guiding principles' and the Government's 'spatial choices' of the structural vision 'Randstad 2040' study (VROM, 2008). For some choices, more detailed information was needed. This was the case with priority number 7, dealing with concentration of urbanisation. The aim of this policy objective was to make maximum use of existing built-up areas in order to preserve open landscapes for agriculture, nature and recreation. VROM used Land Use Scanner to explore the limits of densification and expansion of urban land.

This application encountered two methodological issues. The first issue concerns the *limits to incorporate densification* in the model, specifically the possibility of adding residences and workplaces to an existing built-up area. Land Use Scanner cannot directly analyse the possibilities for *densification* because it only includes a limited number of urban land-use classes, each of which has a fixed density of use and occupation. The projected *densification* would, nevertheless, take place mainly within the same land-use types: single- or multi-story office buildings are, for example, both classified as commercial land use. The model can, however, show the effects of urban *densification* and avoided urbanisation on the preservation of open space. In comparison with a *Baseline scenario* it can show the lack of expansion of new residential and commercial areas.

The second issue concerns the difficulty of determining *limits to expansion* that are socially acceptable. This is important because the demand for (relatively) spacious green urban housing is large, while the supply of space is scarce. Typically, such green urban housing environments have densities of about 17 dwellings per hectare, in contrast with 39–56 dwellings per hectare for innercity urban land use (VROM, 2001). Usually, to preserve open space, various restrictions are imposed that lead to higher dwelling densities and the protection of specific open spaces (Koomen, Dekkers & Van Dijk, 2008a). These limits are incorporated in the model as part of the demand for residential land (through the included dwelling densities) and by specifying (non)suitable locations (through the inclusion of spatial plans and restrictions). The model parameters related to these issues reflect the expected future importance of preserving open space in spatial planning. They imply, therefore, an outcome of the public debate on balancing the degree of individual freedom allowed in choosing a house in a low-density, green urban residential environment and the societal importance of having open spaces. The assumptions related to the dwelling densities included in the model are underpinned with a statistical analysis that is explained further on.

The configuration of the Land Use Scanner based analysis of limits to *densification* and *expansion* is similar to the configuration used in 'Second Sustainability Outlook' study carried out by PBL (MNP, 2007; and Chapter 4 by Rienk Kuiper, Marianne Kuijpers-Linde, and Arno Bouwman, this volume). However, the configuration had to be revised by VROM to meet the specific objectives of the 'Randstad 2040' study. To do so, first a 'Randstad 2040' specific set of valuations of factors for the model suitability maps were derived from the 'Second Sustainability Outlook' study. The so-called Combination Map in that latter study optimises land-use patterns according to a number of policy objectives and describes a potentially sustainable design for spatial development under moderate and high economic and demographic growth rates. The weights assigned to various spatial planning concepts in the Combination Map corresponded well with the policy priorities developed in the 'Randstad 2040' study and were thus maintained. The weights of the suitability maps reflect, for example, a societal preference for concentrated urbanisation. The implementation of this study's policy alternatives focused, therefore, on introducing variations in the demand for residential land to simulate scenario-specific preferences for various residential land-use types.

To explore the spatial impact of the uncertainty in the demand for housing, the *Baseline* and *High Development Pressure scenarios* of 'Second Sustainability Outlook' study were used as they represent a considerable bandwidth of potential urban development. The demand for housing in these growth scenarios is based on demographic projections and assumptions related to, for example, household size, preferred dwelling types, *dwelling densities* and current *densification* rates.

The latter two are especially relevant in relation to the 'Randstad 2040' study and their treatment in the model is discussed here. *Dwelling density* denotes the number of residences per hectare. In the 'Randstad 2040' study, the demand for new residential areas of administrative regions (COROP or NUTS-3) is based on current dwelling densities, derived from clusters of regions with approximately common dwelling densities. These clusters do not necessarily form contiguous regions; however, dwelling densities in the west of the Netherlands (generally included in Cluster 1) are higher on average. See Table 8.1 for an overview of these results.

The table clearly shows the differences between densities used in the 'Second Sustainability Outlook' study and those obtained for the 'Randstad 2040' project. In

Table 8.1 Results of the 'Randstad 2040' analysis of dwelling density (residences per hectare) compared to the 'Second Sustainability Outlook' study. The types of residential areas listed here are defined according to Oskamp, Poulus, and Van Til (2002)

	Second sustainability outlook	Results density analysis			
Type of residential area	National average	National average	Average cluster 1	Average cluster 2	Average cluster 3
Central urban area	43	41	65	40	24
Non-central urban area	38	40	53	37	29
Green urban area	25	23	32	23	18
Village central area	24	19	26	19	16
Countryside	24	13	18	14	11

particular, the new national average dwelling densities for housing in the countryside and villages obtained from Land Use Scanner in the Randstad project are lower than those used previously. This implies that the demand for land for these types of housing increases.

The *densification rate* is considered to be completely dependent on the share within existing urban areas taken by new residences. This share indicates which proportion of new houses built will not claim additional open space. Between 2002 and 2005, around 33% of new dwellings were built within existing urban areas (Snellen, Farjon, Kuiper & Pieterse, 2006) in the Netherlands. In the Randstad conurbation that *densification* rate was lower (26%). Especially in the provinces of Flevoland (which only contains new settlements that were built on reclaimed land since the 1940s) and Zuid-Holland (where, excluding possibilities for *densification* in defunct harbour areas in the Rotterdam area, *densification* may have reached its limits), the share of new dwellings built within existing urban areas was lower. Based on existing, approved municipal zoning plans (NIROV, 2005), the 'Second Sustainability Outlook' study assumes that about 20% of new residential construction will take place in the existing (as of 2000) built-up area for the entire period 2002–2010; much lower rates are expected in subsequent years. Currently, the expected *densification* is approximately 10% for the whole period 2002–2040. In addition to a scenario that simulates land use change with these densification expectations, in the 'Randstad 2040' project a much more ambitious densification rate of 50% of residential land claims has been simulated. An additional measure for saving space was explored by shifting 50% of the demand for green urban and countryside housing to the types central and non-central urban areas. Together we call it the 'ambitious policy' variant in the 'Randstad 2040' study. This ambitious policy variant was applied to the moderate and high growth scenarios and compared to two current policy-based variants, thus generating a total of four scenarios (see Fig. 8.3). The trend-based scenarios are similar to those in the 'Second Sustainability Outlook' study, only differing in the regional division and dwelling densities used.

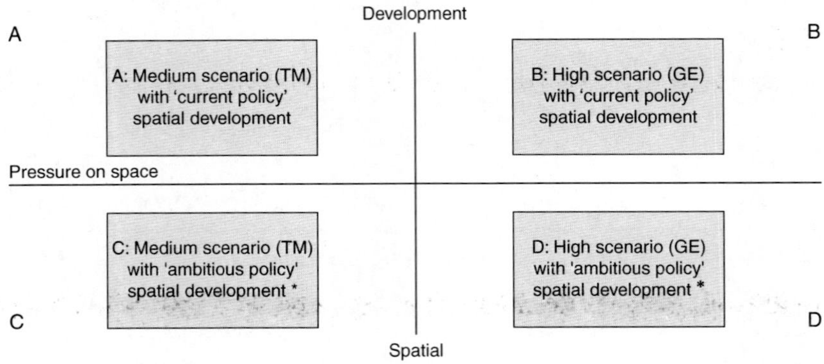

Fig. 8.3 The scenario framework for the 'Randstad 2040' study containing two axes: One indicates variations in ambition level of implemented spatial policies; the other indicates pressure on space. * and 50% intensification and 50% of the demand for green urban and rural housing moved to urban housing

In the 'Randstad 2040' study, the demand for residential land for all four scenarios was specified for an aggregation level that overall distinguishes between 31 regions in the country. This regional aggregation comes from the so-called 'Socrates' model (Poulus and Heida, 2005). This regional division was considered more appropriate than the larger residential regions applied in 'Second Sustainability Outlook' study because the latter study allowed for considerable spatial spread of allocated residential land use. Simplified forms of two of the resulting land-use maps are shown in Fig. 8.4.

8.3.4 Lessons Learned

Land Use Scanner proved to be useful as a research tool for exploring extreme policy scenarios. The *densification* of 50% specified in the 'ambitious policy' variant is an example of such an extreme scenario; the feasibility of this degree of *densification* has not been thoroughly investigated, however. The authorities of the major urban areas in the Randstad conurbation, i.e. the environs of Amsterdam (the so-called 'North wing' of the Randstad) and the environs of Den Haag and Rotterdam (the so-called 'South wing'), also have high ambition levels for densification. The city of Rotterdam even has a target of 80% *densification* for the period 2010–2030. These 'limits of *densification* are thus not easy to assign. The opportunities and possibilities for *densification* depend on many other things. For example, the architectural design of residential areas has some influence on densification options. Achieving maximum *densification* is theoretically possible, but the conditions for maximum densification are unknown, and it is doubtful whether such a degree of densification would be accepted by society. Nevertheless, as the *High Development Pressure scenario* proves, it is clear that if the trend of limited *densification* observed in recent years continues, a large throng of houses will once again be sited on scarce and valuable open space. This has been reality since the end of the Second World

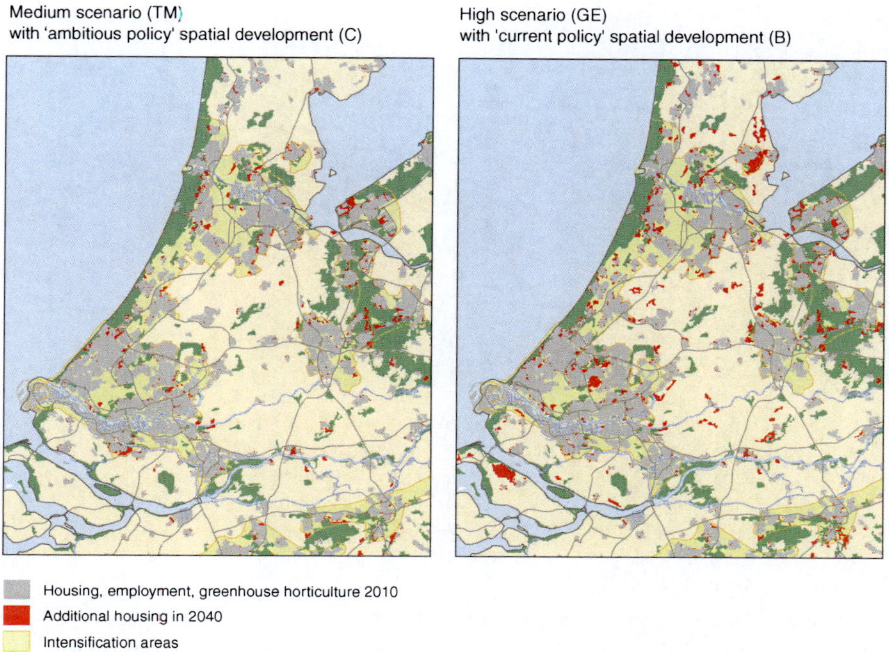

Fig. 8.4 Outcomes of simulations with scenarios C and B of Fig. 8.3 for the 'Randstad 2040' study

War, with the exception of only a few periods when a higher degree of *densification* was actively pursued.

The maps that resulted from simulating urban densification policies are based on assumptions about the proportion of *densification* versus expansion. They show the long-term difference between development under the current policy and development under an ambitious densification policy. The analysis needs to produce the maps facilitates understanding of the playing field in which other relevant interests must be considered. The playing field is presented here in its spatial dimension: a map and the underlying figures on the demand for land. More insight into the limits of *densification* in relation to expansion contributes to making well-founded decisions about a desirable urbanisation strategy for the Randstad conurbation.

8.4 Integrating Climate Adaptation Measures

The Province of Groningen has supplemented their strategic regional plan with an exploration of spatial adaptations to ameliorate the effects of climate change. During a number of workshops the ideas for these spatial adaptations were shared by experts on, for example, energy, ecology, agriculture and climate. Two kinds of workshops were organised to define a coherent set of adaptation measures. *Sector-specific sessions* focussed on generating solutions for problems in a specific domain, while

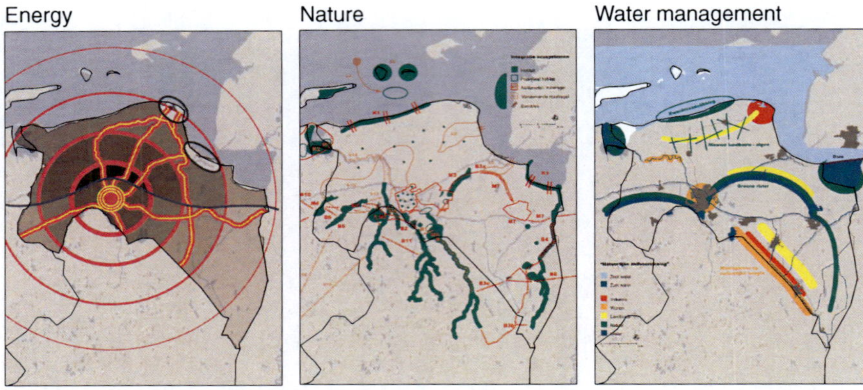

Fig. 8.5 Examples of three climate adaptation measures from the sector-specific sessions in Groningen. *Left*: Potential for alternative energy generation. *Middle*: Proposed extra space to sustain ecosystems. *Right*: Suggested changes in water management

more *general sessions* focussed on establishing an integrated vision for adapting to climate change in Groningen.

In the *sector-specific sessions*, experts proposed measures for adaptations in societal domains such as energy generation, ecology, coastal management and water management. The proposals generated a large number of maps showing threats and proposed counter-measures for a number of specific themes. In a total of four sessions, 28 separate maps were produced, each with a (partially) different set of adaptation measures for Groningen province (For examples of these types of maps, see Fig. 8.5.) These digital maps were subsequently incorporated in a Geographic Information System (GIS).

The *general sessions* used a backcasting approach (Robinson, 1982) as their starting point. In the first session, the participants outlined two visions of climate-change-driven adaptations for Groningen in the year 2100, taking into account different strategies for adapting to a rise in sea level. In the second session, the same groups decided which policies were necessary to establish their visions, thereby creating a third vision. This implementation-oriented session also enabled the participants to rethink and, if necessary, adjust their visions. In the end, three visualisations of a climate-change-proof Groningen were produced and their related policy measures proposed (Van 't Klooster, Pauw & Roggema, 2008).

A second round of general sessions focussed on integrating the results of all preceding sessions. The integration step was initially aimed at clarifying and integrating the sector-specific adaptation measures that differed substantially in spatial and thematic resolution: some maps showed realistic, detailed, local adaptation measures, whereas others only sketched concepts of potential adaptations. To merge these differing views on adaptation into one coherent set of adaptation measures and thereby create a 'climate-proof' Groningen, a GIS-based spatial integration approach was followed (Jacobs, Koomen & Roggema, 2009). In a concluding integration session, participants from the previous workshops confronted the integrated set of climate adaptations with the more visionary backcasting maps to analyse their robustness under different sea-level adaptation strategies (Roggema, 2009).

VU University Amsterdam supported this GIS-based integration process by providing tools, information and analyses of results to the participants of these sessions. GIS makes it easy to share, compare, create and process large quantities of information. All information that was produced during the sector-specific and backcasting sessions was made available to the participants. Furthermore, the results of previous scenario-based Land Use Scanner simulations (documented in Riedijk, Van Wilgenburg, Koomen & Borsboom-van Beurden, 2007) were included in the application. Important steps in the integration approach are: (1) the selection of spatially explicit adaptation measures; (2) a more exact definition of their nature and location; and (3) an analysis of possibilities for combining different adaptation measures.

The first two steps involve a careful analysis of the proposed measures and a conversion into spatial demand for a certain area on the map – a process similar to the configuring a Land Use Scanner simulation. To facilitate the third step, special overlay analysis was performed to assess where demand for land for different adaptation measures potentially overlapped. This was done by dividing the province into grid cells of various resolutions and then counting the number of sector-specific land claims per cell of a specific size. The analysis then showed how many sectors claim land for adaptation measures in a given area (Fig. 8.6). A resolution of 5 km × 5 km was considered to be most effective for providing this information on potential spatial pressure.

In a next step, potential conflicts were identified with these spatial pressure maps. These conflicts were resolved with the help of local and sector-specific experts by either moving or removing specific land claims. This resulted in a map of integrated land claims for adaptation to climate change, which was then overlaid with the three adaptation strategies from the backcasting sessions in the concluding integration session described above.

Simulation results from Land Use Scanner were used to sketch the constraints on adaptation measures posed by current and projected patterns of urbanisation. Human occupation clearly limits the potential for climate adaptation measures at a location: policies that aim to move people, or reclaim built-up land for economically less-intensive land uses are unlikely to be successful. Knowing where people will live and work is, furthermore, necessary to make clear which areas will need strong measures for protection from flooding. Therefore the location of existing and likely future urban areas was added as an additional layer in the overlay analysis of adaptation measures.

A so-called tangible user interface or TouchTable was used in two integration sessions to facilitate interaction between a limited number of participants and the spatial data stored in a computer (Scotta, Pleizier & Scholten, 2006). The interface displays the collected spatial information in a GIS application on a table surface and allows participants to sketch and comment on the projected maps. Anytime a participant touched the table, it registered the specifics of that touch (e.g. place

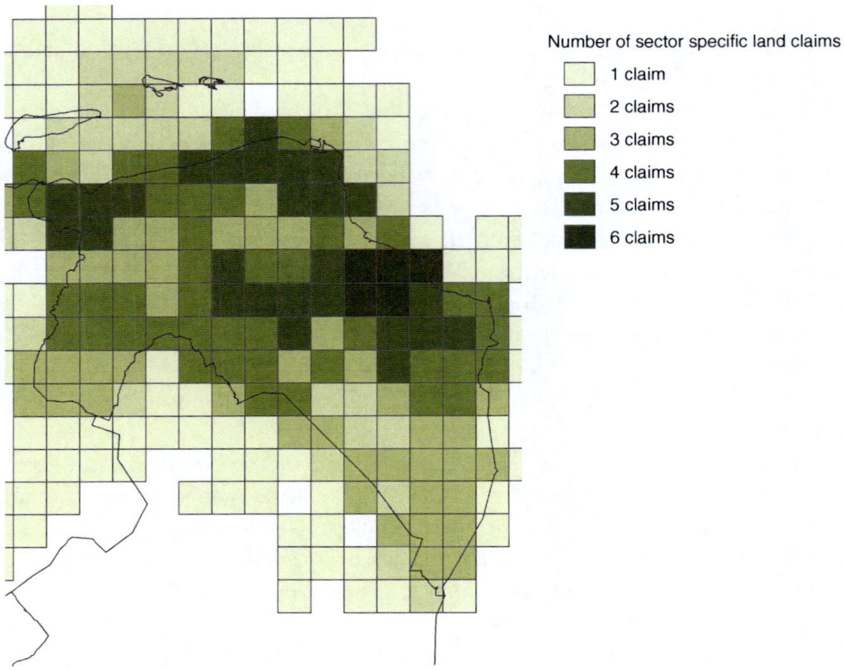

Fig. 8.6 The number of land claims per 5 km × 5 km grid cell following from sector-specific measures for climate-change adaptation and projected future urban areas in Groningen province

on the table, movement, number of elements that were touched). The software on a computer connected to the table then translated the participants' touches into commands to the GIS application.

The organisers found that using a TouchTable in combination with a GIS had a number of advantages over using printed maps or a regular computer screen:

- The table setup stimulated participants to stand around and form a group. It was fun to use and had some 'gadget-value', which enticed people to join in and participate.
- Using a GIS enabled the sharing and viewing of a wide array of spatial information. Furthermore, the GIS facilitated the easy comparison and editing of information and it allowed changes to be reversed and additions made.

8.4.1 Lessons Learned

Several lessons for facilitating multidisciplinary planning are evident from the climate adaptation study for the province of Groningen. First and foremost, the overlay analysis used calls for spatially explicit and non-ambiguous adaptation proposals. This forced participants to reflect on the sector-specific measures proposed in previous sessions. Some of the adaptation measures that were initially proposed were too ambiguous for subsequent analysis. Not all proposals were, for example, explicit in *where* their adaptation measures required land, or in *how much* land was needed. A number of groups proposed multiple scenarios for climate change adaptations, while the proposals of other groups entailed system demands instead of reservations and adaptations of land. Still other groups proposed adaptation measures that took into account all land in the province. Reflection on the previously created sector-specific proposals stimulated participants to make new proposals for various adaptation measures. The integration process thus encouraged reflection on some proposals and eventually led to better definitions of sectoral claims and ideas.

The climate adaptation process for the province of Groningen first focused on stimulating creativity and only later demanded explicit and definite proposals from the many sector-specific groups. This benefit of maximum creativity and flexibility in the sector-specific sessions came at the cost of relatively intensive and time-consuming integration sessions. The process could have been more efficient if the same technology had been used in all sessions.

Using a GIS in combination with a TouchTable and employing overlay analysis appears to be useful for multidisciplinary planning sessions. In the integration sessions, the table was effective in actively involving people in the sessions and the GIS gave relatively easy access to a large number of datasets. The results of the spatial overlay analysis were effective in guiding the participants through possible conflict areas and were very useful in explaining to users the need for being spatially explicit and definite.

Land Use Scanner results were only used to sketch some of the boundary conditions for the development of adaptation measures. A stronger role for Land Use Scanner was considered, for example by using the model to explore the conditions necessary for developing the proposed adaptation measures. This role was not pursued as maximum creativity was preferred in the development of adaptation options. The structured approach of Land Use Scanner in, for example, defining spatial demands and analysing whether they coexist or compete was, however, helpful in structuring the integration sessions. Furthermore, the overlay analysis indicating potential spatial pressure, the extreme flexibility in defining the (location of) adaptation measures, and the application of the TouchTable suggest that the current modelling framework of Land Use Scanner might need to be extended. This project made clear, once again, that quantitative modelling and structured, interactive design can have similar objectives and follow a comparable approach. For providing creative outlooks for the mid-term future, they offer complementary tools, something already demonstrated in an earlier comparative study on Land Use Scanner and design methods (Groen, Koomen, Ritsema & Piek, 2004).

8.5 Parallels

In 2008, Land Use Scanner was used to support a number of regional-planning-related projects, of which several are presented in this chapter. Although the projects described in this chapter varied in their goals and the way the model was used, some parallels in their modelling characteristics, as well as their main purpose, can be observed. A description of these follows in this section. With reference to these parallels, we also discuss here several more general modelling issues related to uncertainty, the supposed success of planning concepts and enhancing the current model layout.

8.5.1 Modelling Characteristics and Main Purpose

An obvious parallel among the modelling characteristics can be observed in the spatial and temporal resolution used while modelling. All projects used a 100 m $\times$ 100 m resolution, and it seems that this scale is sufficient for performing regional studies in a way that is meaningful to planners. The same temporal resolution was also used for all projects. The year 2040 is the furthest into the future in the strategic planning studies discussed in this chapter. Only the Groningen case, with its strong emphasis on climate-driven changes that typically become prominent with a longer time-scale, attempts to look further into the future: as far as 2100. As socio-economic developments are highly uncertain over such long periods of time, land-use simulations were not deemed feasible. Yet, notions of socio-economic change can play an important role in studies on the potential impact of climate change. Initial attempts to create such long-term simulations with Land Use Scanner have therefore been made (Van der Hoeven, Jacobs & Koomen, 2008), but these should first and foremost be considered as philosophising.

Another common characteristic of the land-use simulations we have presented in this chapter is their strong focus on urban development: in all studies, new urban areas are clearly the most important derivative of land-use simulations. Most of the discussions on land-use simulation with the relevant regional authorities centred on the strikingly large demand for residential, commercial and industrial areas.

The main purpose of modelling is normally not to depict the most likely future land-use changes according to observed trends or socio-economic scenarios. Rather, the emphasis is on 'what-if?' types of studies. These studies aim to show what could happen if certain adjustments were to take place. Communicating planning concepts rather than depicting the most likely developments seems, therefore, most important. Furthermore, the intermediate results that were generated in this process proved to be useful ingredients for stimulating discussion and provoking thought about the implications of policy interventions. The process of depiction did, however, appear to be useful for helping to structure and filter some of the issues and interests policy-makers have to deal with.

8.5.2 Modelling Issues

Uncertainty is hardly addressed in the simulations described in this chapter. Apparently, elaborating on a range of possible futures is not desirable in the context of policy-making. Policy-makers seem, in fact, averse to being confronted with uncertainty. They prefer a clear, singular reference point against which policy choices can be tested. Great confidence is therefore placed in the baseline conditions used for 'what-if?' simulations. Several flaws can, however, be observed in these trend-based land-use simulations that, for example, underlie the studies presented in Sections 8.2 and 8.3. Extrapolation of current trends and policies is not necessarily likely to produce plausible results since past changes offer no guarantee for the nature of future changes. Using only one trend obscures the fact that any projection of future socio-economic conditions has a high degree of uncertainty. In fact, the so-called trends are to some extent based on scenario assumptions (e.g. the *High Development Pressure* trend line that is used to model demand for residential land is related to the so-called *Global Economy scenario*). To combine these scenario-based assumptions with other trends that do not follow the same scenario conditions can be considered inconsistent.

Such issues are not very worrisome when simulations are only used to illustrate policy concepts as we have discussed above. However, when land-use simulations have a more substantial role in the formulation process (for example, to highlight potential controversial spatial developments or the impact of policies in Environmental Impact Assessments), the definition of more solid reference points becomes essential. For long-term projections (e.g. 30 years or more) we advocate using a small set of, for example, two to four scenarios covering most of the uncertainty related to key socio-economic conditions to address the full bandwidth of potential land-use changes. To meet policy-makers' need to limit the uncertainty in studies related to concrete planning initiatives for the short term (e.g. 15–20 years), it would be useful to create a well-founded and validated reference scenario that incorporates current trends and policies. Explicit attention should then be paid to underpinning the expected degree of success of the policies included in the scenario.

The what-if type of applications we have presented share a common faith in the supposed success of planning concepts. Planning concepts were generally thought to be fully successful in the land-use simulations we have presented here. Restrictive policies were, for example, quantified as a substantial decrease in suitability of a location for restricted types of land-use, which in most cases prevented the allocation of these land-use types on those locations. The policy goal of intensifying the dwelling densities was also strictly enforced in the simulations of the Randstad project. From rare ex-post evaluations of spatial policies, it is known, however, that such spatial policies are never fully successful. Even the (relatively successful) longstanding Buffer Zones and Green Heart restricted development zones have only slowed down the pace of urbanisation – to about half that of surrounding, non-restricted areas (Koomen et al., 2008a). One should, thus, be very careful in communicating the value of the obtained simulation results. They show what could happen when a certain policy is put in place, but they cannot be used directly to show the likely degree of impact associated with a policy. When simulations are used in the latter type of impact assessments, their implementation in the model (e.g. as a restriction on urban development) should be based on empirical evidence of the degree of success of similar policies in the past. Alternatively, varying degrees of the success of proposed planning concepts can be connected with differing views on the rigidity of planning regimes in different scenarios.

If policy options are not fully restrictive or stimulatory, but rather influence the intensity of land use, their inclusion would benefit from enhancing the current model layout. Densification policies strive to increase the density of occupation, while other policies may strive to limit or reduce the intensity of land use. In the current version of Land Use Scanner, the impacts of such policies cannot be simulated directly. To be able to better assess, for example, the potential impact of varying degrees of prescribed dwelling densities and densification rates on land-use changes, as was done in the Randstad study, ideally the model should endogenously incorporate a notion of land-use intensity that varies with differing degrees of pressure on space. This would make it possible to assess the indirect influence of spatial restrictions on the intensity of urban land use in certain areas. The concept of land-use intensity is closely related to land prices. As the model uses a kind of price mechanism to allocate land, endogenous inclusion of land-use intensity seems a feasible option, albeit a complex one that calls for careful coordination with the external sector-specific models that currently deliver the demands for various types of land.

References

Atzema, O., Van Egmond, K., Mommaas, H., Wenting, R., & Kuijpers-Linde, M. (2008). *Utrecht 2040; Strategische notities in het kader van het traject 'Samen op Weg naar 2040' van de Provincie Utrecht*. Utrecht: Universiteit Utrecht Faculteit Geowetenschappen.

Borsboom-van Beurden, J. A. M., Bakema, A., & Tijbosch, H. (2007). A land-use modelling system for environmental impact assessment; Recent applications of the LUMOS toolbox. Chapter 16. In E. Koomen, J. Stillwell, A. Bakema, & H. J. Scholten (Eds.), *Modelling land-use change; Progress and applications* (pp. 281–296). Dordrecht: Springer.

Bouwman, A. A., Kuiper, R., & Tijbosch, H. (2006) *Ruimtelijke beelden voor Zuid-Holland*. Rapportnummer 500074002.2006. Bilthoven: Milieu- en Natuurplanbureau.

Cramer, J. (2008). *Nieuwe maten van de Randstad. Nova Terra [Januari 2008], 4–7*. NIROV: The Hague.

Groen, J., Koomen, E., Ritsema, van E. J., & Piek, M. (2004) *Scenario's in kaart; model- en ontwerpbenaderingen voor toekomstig ruimtegebruik*. Rotterdam/Den Haag: NAi Uitgevers/Ruimtelijk Planbureau.

Jacobs, C. G. W., Koomen, E., & Roggema, R. (2009). *Towards an integrated vision of a climate proof Groningen (forthcoming)*. Groningen: Provincie Groningen.

Koomen, E., Dekkers, J., & Van Dijk, T. (2008a). Open space preservation in the Netherlands: Planning, practice and prospects. *Land Use Policy, 25*(3), 361–377.

Koomen, E., Kuijpers-Linde, M., & Loonen, W. (2008b). *Ruimtelijke verkenning Overijssel 2040*. Amsterdam: Geodan Next.

Kuijpers-Linde, M., Koekoek, A., & Loonen, W. (2008). *Uitwerking ruimtelijke beelden voor het nieuwe omgevingsbeleid van Drenthe*. Amsterdam: Werknotitie. Geodan Next.

MNP (2007). Nederland Later; Tweede *Duurzaamheidsverkenning deel fysieke leefomgeving Nederland*. MNP-publicatienr.500127001/2007. Milieu- en Natuurplanbureau, Bilthoven.

NIROV (2005). *Nieuwe Kaart, Nieuwe Ruimte: Plannen voor Nederland in 2015*. Den Haag: Nirov.

Oskamp, A., Poulus, C., & Van Til, R.-J. (2002). Spatial concentration and deconcentration of household types in the Amsterdam region; Effects of three scenarios of new construction. *Journal of Housing and the Built Environment, 17*, 321–335.

Poulus, C., & Heida, H. (2005). *Methodiek en toelichting Socratesmodel*. Delft: ABF Research.

Riedijk, A., Van Wilgenburg, R., Koomen, E., & Borsboom-van Beurden, J. (2007). *Integrated scenarios of socio-economic change. SPINlab research memorandum SL-06*. Amsterdam: Vrije Universiteit Amsterdam.

Robinson, J. (1982). Energy backcasting: A proposed method of policy analysis. *Energy Policy, 10*, 233–337.

Roggema, R. (2009). *Hotspot klimaatbestendig Groningen: Eindrapportage ontwerp*. Groningen: Provincie Groningen.

Scotta, A., Pleizier, I. D., & Scholten, H. J. (2006, May 15–17) Tangible user interfaces in order to improve collaborative interactions and decision making. In E. Fendel & R. Rumor (Eds.), *Proceedings of 25th Urban Data Management Symposium* (*UDMS 2006*). Aalborg, Denmark: Delft, Urban Data Management Society.

Snellen, D., Farjon, H., Kuiper, R., & Pieterse, N. (2006). *Monitor nota ruimte. De opgave in beeld*. Den Haag/Bilthoven: Ruimtelijk Planbureau/Milieu- en Natuurplanbureau.

Van der Hoeven, E., Jacobs, C. G. W., & Koomen, E. (2008). *Beknopte beschrijving van sociaaleconomische scenario's voor het jaar 2100*. Amsterdam: Vrije Universiteit.

Van 't Klooster, S., Pauw, P., & Roggema, R. (2008). *Backcasting analyse van een klimaatbestending Groningen*. Groningen: Provincie Groningen.

VROM (2001). *Ruimte maken, ruimte delen. Vijfde nota over de ruimtelijke ordening 2000/2020*. Den Haag: Ministerie van Volkshuisvesting Ruimtelijke Ordening en Milieubeheer.

VROM (2008). *Structuurvisie Randstad 2040. Naar een duurzame en concurrerende Europese topregio*. Den Haag: Ministerie van Volkshuisvesting Ruimtelijke Ordening en Milieubeheer.

Part III
Future Developments

Chapter 9
Explaining Land-Use Transition in a Segmented Land Market

Potential Input for Land Use Scanner

Jasper Dekkers and Piet Rietveld

9.1 Introduction

Spatial planning policies can create segmented sub-markets for land, leading to an artificial scarcity for certain types of land use (usually urban land use) and higher land prices in that submarket, which can eventually spillover to other sub-markets. In a spatial context, these spillover effects are mainly observed in the urban fringes, where rural land owners often ask a higher price for their rural land because of expectations that this will be converted into urban land. The price for rural parcels in the urban fringe is thus related to the probability that these parcels will be put to urban uses in the (near) future, which is referred to as the 'transition probability'. Figure 9.1 describes this relation in more detail. Land prices decline as the distance from the city centre increases. In situations where there are no spatial planning constraints (Fig. 9.1a), there is no gap between land prices at the urban fringe. If there is a spatial planning policy being implemented that restricts building on the fringe of a city (Fig. 9.1b), a sudden change in land prices can be observed for rural parcels immediately bordering the city. When it is expected that these limitations along the fringe may be lifted, the market price for those rural parcels will be somewhere between the urban and the agricultural land price. If the market price is closer to the agricultural land price, there is a low transition probability, and if the market price is closer to the urban land price, there is a high transition probability.

Various approaches exist for analysing land markets. One approach is to study land market *processes* in a descriptive way – this is commonly referred to as institutional analysis (e.g. Segeren, Needham & Groen, 2005; Segeren, 2007). Another, very different approach is to study land market *outcomes* in a quantitative way. In this chapter we use this second approach, since previous studies (e.g. Buurman, 2003; Dekkers, Rietveld, Van den Brink & Scholten, 2004) have already confirmed that quantitative models such as the Hedonic Pricing Method (Rosen, 1974) can effectively incorporate multiple sub-markets at the same time. We do

---

J. Dekkers (✉)
Department of Spatial Economics/SPINlab, VU University Amsterdam, De Boelelaan 1105, 1081 HV Amsterdam, The Netherlands
e-mail: j.dekkers@vu.nl

E. Koomen, J. Borsboom-van Beurden (eds.), *Land-Use Modelling in Planning Practice*, The GeoJournal Library 101, DOI 10.1007/978-94-007-1822-7_9,

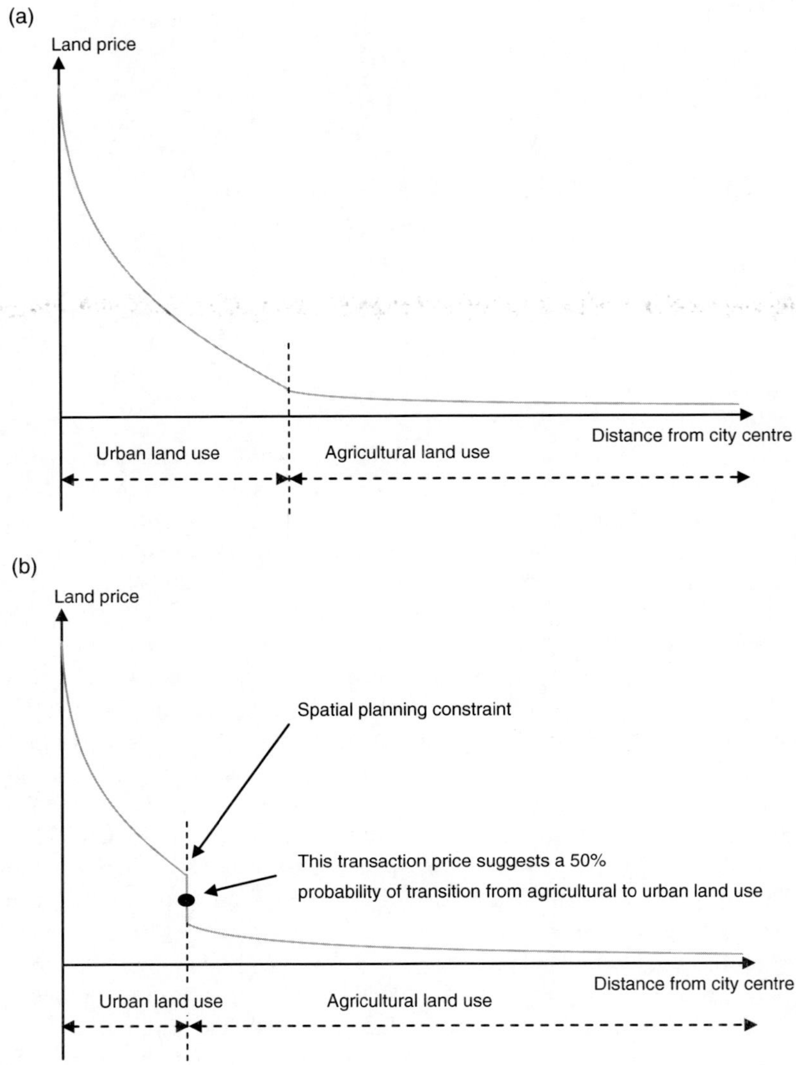

Fig. 9.1 Relation between land price, land use and distance from city centre: (**a**) with no spatial planning constraints; (**b**) with spatial planning constraints (Source: Dekkers, 2010)

think, however, that an institutional analysis of the land market could be done to complement the approach we use here, since it can improve the understanding of the driving forces and motivations of (groups of) actors on various sub-markets and provide insights into the processes in these markets that lead to the market outcomes we are studying.

In this chapter we present a novel analytical method for explaining how probable locations for future urban development can be identified from information on actual land prices. In Section 9.2, we describe our approach for explaining the relation between transition probabilities and land prices using a single-step, multiple-equation approach. In Section 9.3, we apply this model to the land market in the Province of Noord-Holland and discuss the results. In Section 9.4, we use the results from the model to derive a transition probability map for the entire province, thus including land that has not been traded. There we also discuss how our method can be used in combination with land-use models, such as the Land Use Scanner model, to support spatial planning processes in The Netherlands. Finally, in Section 9.5 we offer some concluding remarks.

9.2 A New Explanatory Model for Land-Use Transitions

Our new method for modelling land-use transitions integrates the strong points of two existing quantitative models for explaining land prices in The Netherlands (Buurman, 2003; Luijt, Kuhlman & Pilkes, 2003) into a linear probability model, thus combining Buurman's large set of explanatory variables and the probabilistic approach of Luijt et al.

Buurman (2003) and Luijt et al. (2003) both use a single-equation logarithmic regression model for explaining land prices. Their models are revealed-preference models that use the Hedonic Pricing Method (HPM). The HPM determines the *implicit* value of non-tradable characteristics of goods by analysing the *observed* values of tradable goods that incorporate all or part of those non-tradable characteristics. Rosen (1974) was the first to publish a generally accepted article about this method. For a more in-depth summary of the hedonic pricing technique, we refer interested readers to Griliches (1971, Chapter 1) and Gordon (1990).

The models of Buurman and Luijt et al. both use cadastral data on transactions for parcels of land outside urban areas. And both include numerous factors – based on extensive literature research – that are most likely to affect transaction prices of land. These factors can be divided into *transaction characteristics* (e.g. price, date of sale, type of buyer, type of seller), *parcel characteristics* (e.g. parcel size, soil quality) and *spatial characteristics* (e.g. accessibility measures, environmental measures, zoning designations).

Both models have the drawback that not all characteristics are equally important for all buyers. For instance, project developers most likely do not value soil quality as much as farmers do. Therefore, Luijt et al. (2003) proposed an alternative for the single-equation model in the form of a two-step, multiple equation model (Fig. 9.2), using separate explanatory models for 'red'(i.e. urban) and 'green' (i.e. agricultural) buyers. In order to determine which equation to use for a transaction, the transition probability must first be estimated using parcel and buyer characteristics. Luijt et al. (2003) discerned building constructors, municipalities and project developers as red buyers and farmers as green buyers. On the vertical axis of Fig. 9.2, $price_{gi}$ is the green transaction price and $price_{ri}$ is the red transaction price. The horizontal axis shows the transition probability. Logically, the higher the transition probability from green to red of a certain parcel, the higher the price of that parcel.

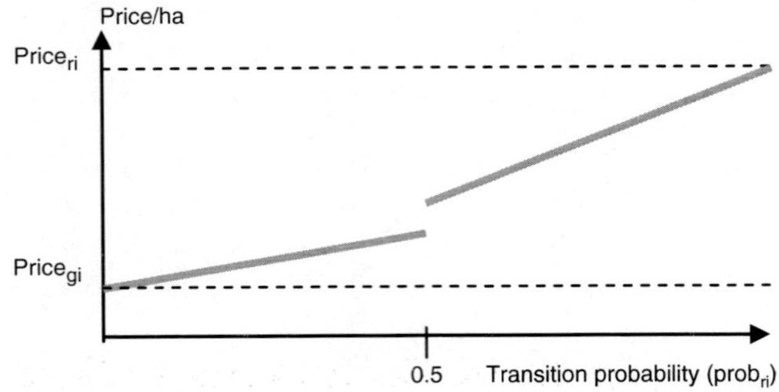

Fig. 9.2 Transition probability model of Luijt et al. (2003) using a two-step multiple equation (Source: Dekkers, 2010)

A major drawback of using an artificial cut-off value at $prob_{ri} = 0.5$ is the uncertainty for land transactions with probabilities around this value of being allocated to the correct equation, thus reducing the explanatory power of the model around this point. In the case study of Luijt et al. (2003), the results of the estimations show that the two functions certainly do not connect.

Drawing on the two-step, probabilistic approach of Luijt et al. and the richness of the explanatory factors in Buurman's approach, we have developed a new modelling approach for:

- estimating hedonic price models for the green and the red sub-markets separately, in order to do justice to the segmented nature of land markets;
- computing transition probabilities from green to red in order to describe market expectations concerning observed transitions;
- determining the contribution of explanatory variables to these transition probabilities;
- using this model to predict transition probabilities for zones where no transactions have taken place.

The analytical steps necessary to obtain the desired results with this approach are summarised in Fig. 9.3.

A more detailed description of our approach now follows. Suppose a potential buyer foresees a transition probability from green to red of $prob_{ri}$. Suppose, also, that we know the prices of the parcels of land concerned if they belong in the green market segment ($price_{gi}$) and their prices if they belong in the red segment ($price_{ri}$). How much is the buyer willing to offer? This model can be depicted graphically as in Fig. 9.4. Note that in this model the market price is equal to the green price when the transition probability equals 0. The market price equals the red price when the transition probability is 1. Intermediate market prices are related to intermediate probabilities.

Step 1a	Find subset of transactions for which it is clear on an a priori basis that the new use of the parcels concerned is red. These are called "red transactions".
Step 1b	Same for green transactions. This leads to a set of "green transactions".
Step 1c	The remaining transactions are called "uncertain transactions".
Step 2a	Estimate hedonic price model for subset of red transactions.
Step 2b	Same for green transactions.
Step 3a	Use the red price model estimated in Step 2a to predict the value that all parcels (red, green, uncertain) would have according to red market conditions. This leads to predicted red prices for all parcels.
Step 3b	Same for green price model. This leads to predicted green prices for all parcels.
Step 4	Compute for all transactions the transition (from green to red) probability as probri as the ratio of the difference between the actual price minus the predicted green price divided by the difference between the predicted red price and the predicted green price.
Step 5	Use a linear probability model to explain the values of the transition probability defined in Step 4 by means of a set of explanatory variables.
Step 6	Compute the transition probability of all zones in the region concerned, also including the zones where no transactions took place.

Fig. 9.3 Summary of steps in linear probability approach (Source: Dekkers, 2010)

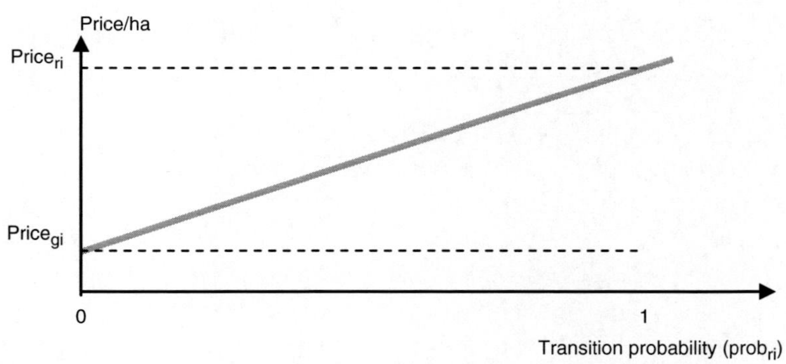

Fig. 9.4 Single-equation transition probability model (Source: Dekkers, 2010)

Step 1a – Find red subset of transactions

In order to apply this method we need estimates of red and green prices. To do this, we can select a subsample of transactions for which we know that the parcels concerned will almost certainly be developed, i.e. $prob_{ri} = 1$. In this case, we choose to take a sample of 10% of the transactions with the highest price per square metre.

Step 1b – Find green subset of transactions

Similarly to Step 1a, consider a sample of transactions where, according to expert judgement, $prob_{ri}$ is very close to zero: for example, a parcel lying within the

National Ecological Network and for which, in that area, another policy applies that restricts urbanisation. As with the red subsample, we choose to take a sample of 10% of the transactions with the lowest price per square metre.

Step 1c – Remaining transactions

The remaining 80% of the transactions are only used from Step 3 onwards.

Steps 2a & 2b – Estimate hedonic model for red and green transactions

Since for the red sample selected in Step 1a we assume that $prob_{ri} = 1$, this means we can estimate for the red transactions:

$$price_i = \alpha_r + \beta_r \cdot x_{ri} + \varepsilon_i \tag{9.1}$$

where α_r is a constant and β_r is a vector of coefficients for the related red matrix of independent variables (x_{ri}). This leads to estimates $\hat{\alpha}_r$and $\hat{\beta}_r$.

In the case of the green transactions, we assume that $prob_{ri} = 0$, so we can again simplify the basic equation and estimate:

$$price_i = \alpha_g + \beta_g \cdot x_{gi} + \varepsilon_i \tag{9.2}$$

where α_g is a constant and β_g is a vector of coefficients for the related green matrix of independent variables (x_{gi}). This leads to estimates $\hat{\alpha}_g$and $\hat{\beta}_g$.

Steps 3a & 3b – Compute red and green transaction prices

Once we know $\hat{\alpha}_r$and $\hat{\beta}_r$ and $\hat{\alpha}_g$ and $\hat{\beta}_g$, we can compute red and green prices for all transactions:

$$\hat{price}_{ri} = \hat{\alpha}_r + \hat{\beta}_r \cdot x_{ri} \text{ and } \hat{price}_{gi} = \hat{\alpha}_g + \hat{\beta}_g \cdot x_{gi} \tag{9.3}$$

In this way, we can obtain predictions of land prices for all parcels concerned, albeit conditional on whether they fall under the red or the green regime.

Step 4 – Compute transition probability

We can now compute the transition probability for all transactions based on the model represented in Fig. 9.4:

$$price_i = (1 - prob_{ri}) \cdot \hat{price}_{gi} + prob_{ri} \cdot \hat{price}_{ri_i} \tag{9.4}$$

After rewriting this formula, we can compute $prob_{ri}$ using $price_i$ and the predicted red and green land prices $\hat{price}_{ri}$ and $\hat{price}_{gi}$:

$$prob_{ri} = \frac{\hat{price}_{\mathrm{gi}} - price_{\mathrm{i}}}{\hat{price}_{\mathrm{gi}} - \hat{price}_{\mathrm{ri}}} \tag{9.5}$$

Step 5 – Explain transition probability

Next, we can estimate the factors that have an impact on $prob_{ri}$. Let these factors be denoted as z_i (mainly overlapping with x_i, only the surface and time dummy variables are excluded). We have to take into account that there will probably be observations for which the market price lies either below the estimated green price or above the estimated red price. Plausible causes for prices dropping below the green price could be, for instance, unobserved features such as soil contamination. Unobserved features could also explain why an observed price is higher than the red, urban price.

A linear probability model is used to explain variations in the transition probability $prob_{ri}$:

$$prob_{ri} = \gamma + \delta \cdot z_i + \mu_i \tag{9.6}$$

where γ and δ are parameters to be estimated, and μ_i is an error term.

Step 6 – Compute transition probability for all zones

In this final step we can use the coefficients of the linear probability model in combination with the underlying spatial data sets of the explanatory variables to calculate the transition probability for all zones in the region concerned. This can be done by rasterising all data sets on, for instance, a 25 m grid.

9.3 A Case-Study in the Province of Noord-Holland

9.3.1 Description of the Study Area

The land market in Noord-Holland (Fig. 9.5) consists of many sub-markets. There are many agricultural areas for which various nature protection policies and/or nature development plans apply. Also, a great deal of land is used for horticulture and flower bulb cultivation, which is relatively expensive. Then there is the national airport, Schiphol Amsterdam airport, which is assumed to have a large impact on land use and prices in its surrounding areas. In our opinion, the very heterogeneity of land uses and prices was an interesting setting in which to apply our model (as described in Section 9.2).

The surface area of the Province of Noord-Holland covers 4,059 km^2, of which 2,657 km^2 is dry land. This equals 7.8% of the land surface of the Netherlands. In 2000, 2.5 million people were living in Noord-Holland, making it the province with the second highest number of inhabitants in the Netherlands. Approximately 19% of the national Gross Domestic Product (GDP) is generated in Noord-Holland (CBS, 2000). Table 9.1 shows that the 'Commercial services' sector is relatively important in Noord-Holland and that the 'Manufacturing' and 'Agriculture, forestry and fisheries' sectors are relatively less important. As in the rest of the Netherlands, agricultural land use is decreasing in Noord-Holland, while urban and natural land use are increasing. Compared to the Netherlands as a whole, Noord-Holland has less forest and nature and more built-up areas.

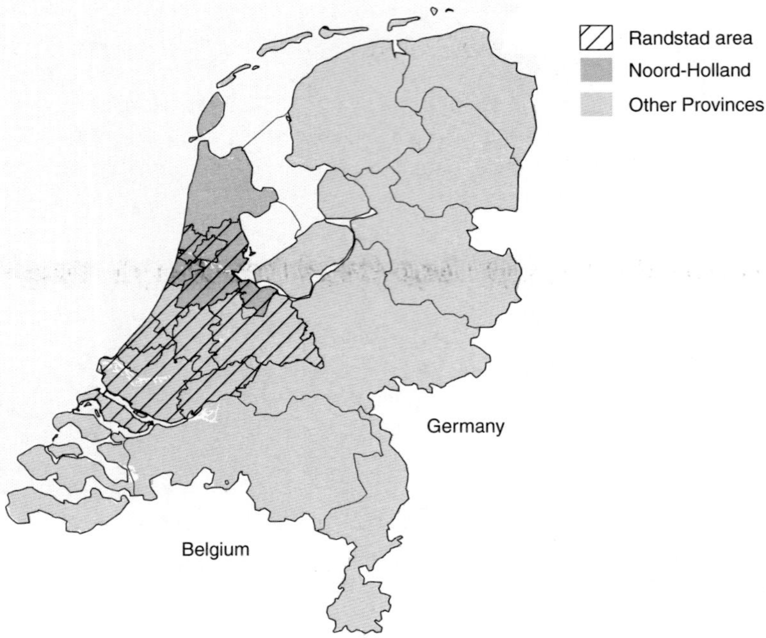

Fig. 9.5 Location of the Province of Noord-Holland within The Netherlands (Source: Dekkers, 2010)

Table 9.1 Relative sectoral contributions to the economies of Noord-Holland and The Netherlands (Source: CBS, 2000)

	Gross added value 1999 (%)	
Sector	Noord-Holland	The Netherlands
Agriculture, forestry and fisheries	1.6	2.8
Industry	17.0	25.5
Commercial services	59.3	49.0
Government and healthcare	22.2	22.8
Total	100.0	100.0

9.3.2 Description of the Land Transaction Data

For our case-study, we used the cadastral InfoGroMa database of DLG (Government Service for Land and Water Management, part of the Ministry of Agriculture, Nature and Food Quality). This data set is a subset of the Dutch cadastral database containing all transactions for parcels outside urban areas in the Netherlands.

When we disregarded transactions within families, the database registered for Noord-Holland 2,685 parcels in 1,625 transactions for the period 1998–2002. We also disregard transactions containing immobile property, as we do not have information on the characteristics of the immobile property and the influence of these characteristics on the transaction price.

Table 9.2 gives some descriptive statistics for both the red and green subsamples (Steps 1a and 1b). We see large differences in the transaction price per square metre

Table 9.2 Summary statistics for the red and the green subsamples related to Step 1a and 1b

Variable	R (red) G (green)	Min.	Max.	Mean	Std. dev.
Transaction characteristics					
Transaction price (× 1,000 euros)	R	15.24	29,743	857.31	2,583
	G	1.53	267.69	30.84	38.45
Transaction price per square metre	R	5.15	153.15	19.83	20.14
(euros/m^2)	G	0.16	1.07	0.55	0.19
Leased land (0/1)	R	0.00	0.00	0.00	0.00
	G	0.00	1.00	0.54	0.50
Structural characteristics					
Surface area (× 1,000 m^2)	R	2.50	592.34	45.01	80.94
	G	2.51	569.20	56.25	73.82
Spatial characteristics					
Share of a transaction that is located in	R	0.00	1.00	0.33	0.47
an urban development zone (New Map	G	0.00	1.00	0.01	0.08
of The Netherlands)					
Distance to the Randstad (km)	R	0.00	77.29	15.00	17.56
	G	0.00	80.09	23.54	19.99
Distance to the nearest built-up area (km)	R	0.00	1.81	0.29	0.28
	G	0.00	2.23	0.57	0.42
Distance to main road < 200 m (0/1)	R	0.00	1.00	0.13	0.34
	G	0.00	1.00	0.07	0.26
Share of a transaction's land use that is	R	0.00	1.00	0.06	0.24
urban green	G	0.00	1.00	0.01	0.11
Share of a transaction's land use that is	R	0.00	1.00	0.05	0.21
greenhouse horticulture	G	0.00	0.00	0.00	0.00
Land development project (BBL/DLG)	R	0.00	1.00	0.23	0.42
(0/1)	G	0.00	1.00	0.40	0.49
Share of a transaction that is located in	R	0.00	1.00	0.25	0.43
the provincial ecological main structure	G	0.00	1.00	0.65	0.47
(PEHS)					
Share of a transaction that is located in a	R	0.00	1.00	0.10	0.30
Buffer Zone	G	0.00	1.00	0.20	0.40
Share of a transaction that is located in a	R	0.00	1.00	0.07	0.26
belvedere policy zone (UNESCO world	G	0.00	1.00	0.08	0.26
heritage area)					
Share of a transaction's land use that is	R	0.00	1.00	0.01	0.11
nature	G	0.00	1.00	0.06	0.24

Note: Transactions in Noord-Holland (period 1998–2002; $N = 1{,}625$; $N_{red} = 163$; $N_{green} = 162$)

between the two subsamples. The mean price per square metre of the red subsample is about 36 times higher than that of the green subsample. This underlines the strong segmentation between the green and the red land markets in the Netherlands. No leased land was found in the red subsample and, on average, transactions in the red subsample lay closer to the Randstad and to urban areas, often within 200 m of a main road. Further, in the red subsample, on average a larger share of the transactions lay within and were subject to urban development plans or were classified as 'urban green' or 'greenhouse horticulture' land-use types. Similar to the distinction between land uses as made in Koomen, Dekkers, and Van Dijk (2008), the seemingly contradictory urban green category contains sparse vegetation (bare soil or grassland) in or nearby urban areas. It contains land with a predominantly green appearance and a functional relation to the neighbouring urban area, such as sports-fields, parks and land set aside for construction. Greenhouses are distinguished separately from other agriculture because of their distinct urban appearance.

In contrast, in the green subsample, on average a larger share of the transactions lay within the National Ecological Network. The same applies for Buffer Zones (green corridors no less than 4 km wide that must be left open between major agglomerations; RNP, 1958) and for the land-use type 'nature'. For Belvedere policy zones (UNESCO world heritage areas; OCW, LNV, VROM & V&W, 1999), there is virtually no difference between the two subsamples. Figure 9.6 shows four maps containing the locations of the sold parcels for the two subsamples, family transactions, and the remaining parcels; recall that we disregarded family transactions in our analysis. Amsterdam is located at the bottom of the maps (the harbour area is just visible). What is clearly visible is that the red parcels (Fig. 9.6b) are in general concentrated near urban areas, while the green parcels (Fig. 9.6c) are generally further away from urban areas.

9.3.3 Selecting Model Variables

The variables we used overlap to some extent with the variables we used in our explanatory Hedonic Pricing model for the land market in the Province of Noord-Holland, which is described in Dekkers et al. (2004). That model is the same as the one Buurman (2003) used. The difference is that our current analysis does not exclusively focus on the explanation of rural land prices but rather on rural-urban transition probability. For that reason, we included factors that we expected would have an influence on this transition probability. First, we included transport noise as a factor in the red explanatory Hedonic Pricing model by defining a dummy variable that had the value 1 when a parcel is located less than 200 m from a main road. We also included information on land use: built-up, urban green and greenhouse horticulture in the red model; and nature in the green model. Finally, in the green explanatory model we took into account whether or not parcels lay in a Buffer Zone or a Belvedere/UNESCO Zone. We expected these policies to negatively influence

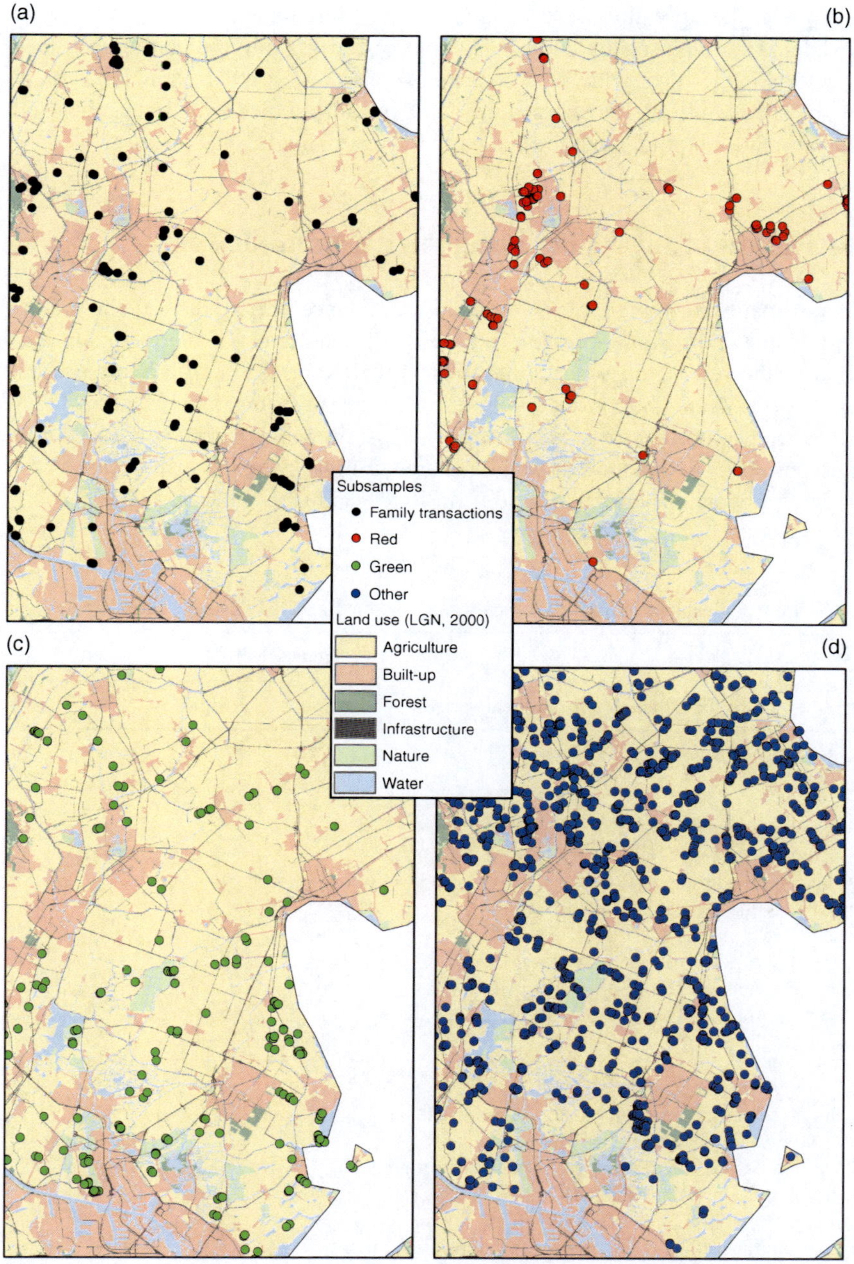

Fig. 9.6 Locations of all geocoded rural land transactions without immobile property in a part of the Province of Noord-Holland in the period 1998–2002 and land use as in 2000 (Source: Dekkers, 2010)

land prices because they restrict land use to some extent. After testing various model and variable specifications, we chose a semi-logarithmic model specification since this is the most widely used form. We also tested other variables related to, for instance, soil quality, accessibility and spatial policies aimed at nature and/or recreation development, but these factors were not significant.

9.4 Results of the Model Parameter Estimations

The estimation results (Table 9.3) give us confidence in both the red and the green models: the coefficients show the expected signs and most of them are significant. First, in the red model we see that parcels that were closer to urban areas had a higher price. Next, parcels that were less than 200 m from a main road had a lower price than parcels located further away, the most probable reason being the nuisance experienced from transport noise in this area. Further, when the land use was urban green or greenhouse horticulture, this positively influenced parcel prices.

Table 9.3 Results for the subsample estimates in the formulas related to Step 2a and 2b

	Red ($N = 162$)		Green ($N = 162$)	
Variable (LN(price-deflated) = dependent variable)	Coeff.	Sign.	Coeff.	Sign.
Constant	2.426	***	−1.289	***
LN(surface)	0.972	***	1.021	***
Share of a transaction that is located in an urban development zone (New Map of The Netherlands)	0.189	*		
Distance to the Randstad (km)	−0.002			
Distance to the nearest built-up area (km)	−0.255			
Distance to main road < 200 m	−0.256	*		
Share of a transaction's land use that is urban green	0.397	*		
Share of a transaction's land use that is greenhouse horticulture	−0.085			
Leased land			−0.116	**
Land development project (BBL/DLG)			−0.045	
Share of a transaction that is located in the provincial ecological main structure (PEHS)			−0.078	
Share of a transaction that is located in a buffer zone			−0.125	*
Share of a transaction that is located in a Belvedere policy zone (UNESCO world heritage area)			−0.028	
Share of a transaction's land use that is nature			−0.097	
s.e. of regression coefficient	0.589		0.303	
R^2	0.797		0.928	
Adjusted R^2	0.788		0.924	

Note: *** = significant at 0.01; ** = significant at 0.05; * = significant at 0.10

In the green model, we see that leased land was cheaper, as was land that was in a land development project area of DLG or in the provincial concretisation of the National Ecological Network. Two major reasons why DLG buys land are its use in re-allotment projects and to create new nature areas. In both cases, the land use remains green: agriculture or nature, which certainly does not have a positive impact on prices. For parcels in Buffer Zones, we do indeed see a negative influence on the land price, confirming our expectation. The Buffer Zone policy has a strong protective value (Koomen et al., 2008; Van Rij, Dekkers & Koomen, 2008). When Buffer Zones were established in the 1960s, the original intent was that agriculture would be a driving force in these areas, in combination with nature and recreational developments. Over the years, the national government has changed its strategy more and more toward actively acquiring land for nature development. Apart from that, Buffer Zones are, in general, not very large areas, and they are to be found in areas between major urban agglomerations where urbanisation pressures are already high. For all these reasons, scale increases for agricultural businesses are hard to realise, so that these areas have less potential for modern farming than elsewhere. The same kind of reasoning applies to the Belvedere/UNESCO areas: this policy also has a highly protective value, as it is difficult to expand farming activities here and multiple restrictions apply. The model coefficient of the Belvedere policy zone confirmed our expectations as it had a negative sign, but this variable is not significant.

We subsequently used the coefficients of the red and the green model to estimate red and green transaction prices and to compute the rural-urban transition probability. Figure 9.7a and b show the estimated prices (x-axis) plotted against

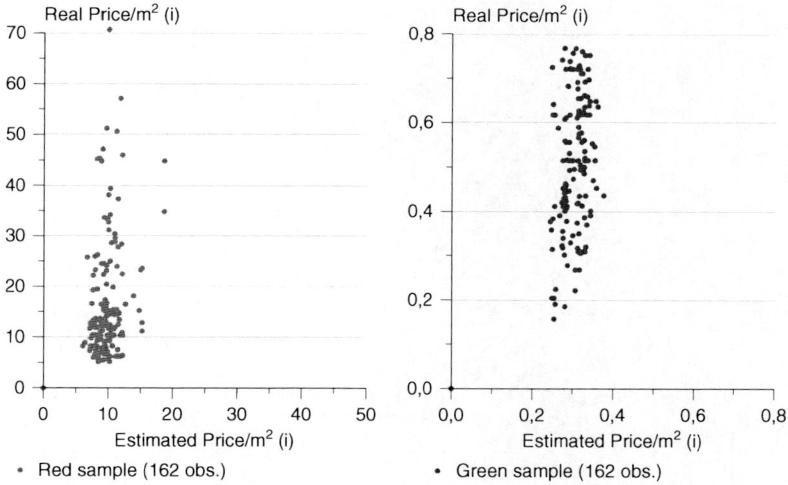

Fig. 9.7 Plot of real versus estimated transaction price per m^2 for the red and green sub-samples, respectively (some outliers are not shown) (Source: Dekkers, 2010)

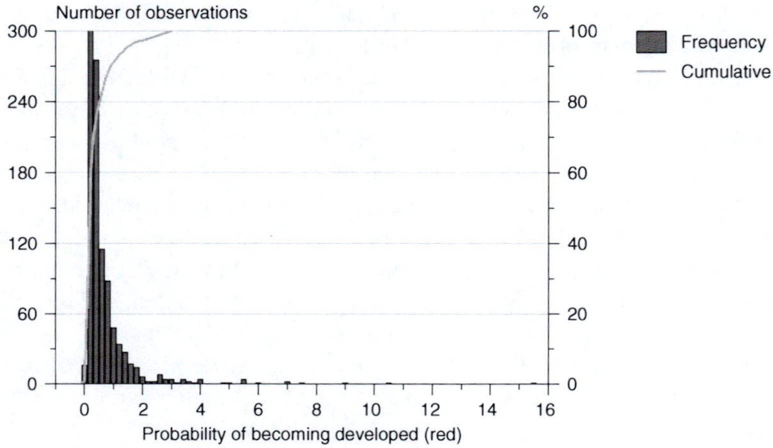

Fig. 9.8 Histogram of calculated rural-urban transition probabilities of the total sample (Source: Dekkers, 2010)

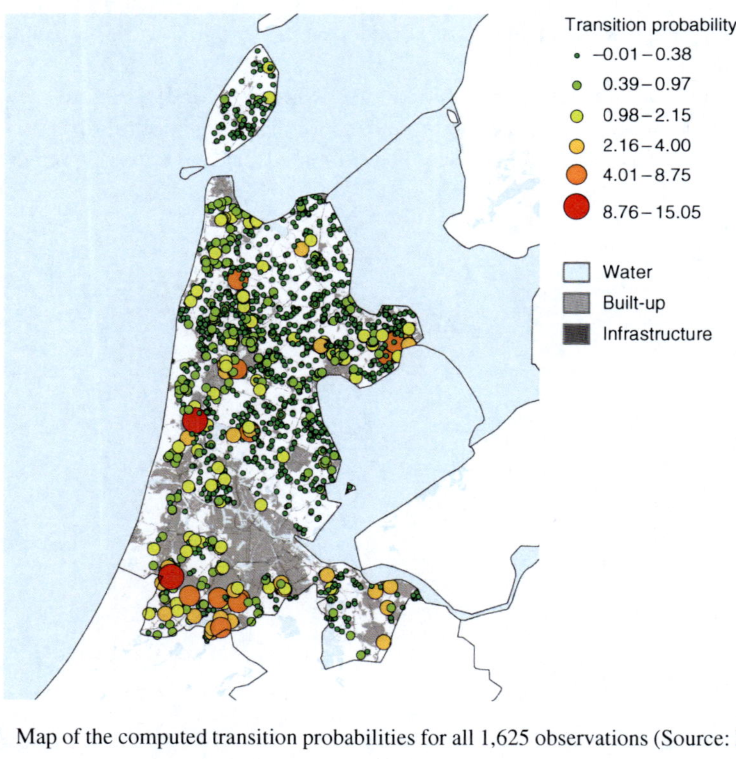

Fig. 9.9 Map of the computed transition probabilities for all 1,625 observations (Source: Dekkers, 2010)

the transaction prices (y-axis) for red and green transactions, respectively. Both models somewhat underestimated land prices and in particular the red model does not capture the full variation in land prices.

As Fig. 9.8 shows, more than 90% of the calculated transition probabilities lie between 0 and 1, meaning that the Ordinary Least Squares (OLS) estimation does not have to correct for a large number of probabilities outside this range: there are no probabilities below 0 and only 8.8% of the observation's probabilities are above 1. Theoretically, probabilities cannot have values outside the range [0:1]. In practice, however, these values can occur due to unobserved features in the model (see also Step 5 in Section 9.2). The mean probability is 0.40 and the range is [0.01:15.05]. As can be seen, there are some observations with a very high transition probability (see also Fig. 9.9). The explanation is that in the areas where these transactions have taken place, a lot of urban development has occurred, with numerous 'red' factors/variables simultaneously having a positive influence on the transition probability.

The results of the OLS analysis shown in Table 9.4, with the rural-urban transition probability as the dependent variable, indicate that the coefficients of the 'red variables' explain these variables well: when a parcel lies in an area with

Table 9.4 Results for the ordinary least squares estimation on all 1,625 observations according to Eq. (9.6)

	OLS results	
Variable	Coeff.	Sign.
Constant	0.773	***
Urban development plans (New Map of The Netherlands)	0.724	***
Distance to the Randstad (km)	−0.008	***
Distance to the nearest built-up area (km)	−0.104	**
Distance to main road < 200 m	0.044	
Share of a transaction's land use that is urban green	0.051	
Share of a transaction's land use that is greenhouse horticulture	0.949	***
Leased land	−0.257	***
Land development project (BBL/DLG)	−0.137	***
Share of a transaction that is located in the provincial ecological main structure (PEHS)	−0.093	**
Share of a transaction that is located in a buffer zone	−0.226	***
Share of a transaction that is located in a Belvedere policy zone (UNESCO world heritage area)	−0.262	***
Share of a transaction's land use that is nature	−0.125	
s.e. of regression coefficient	0.766	
R^2	0.142	
Adjusted R^2	0.135	

Note: *** = significant at 0.01; ** = significant at 0.05; * = significant at 0.10

urban development plans, the probability of a green parcel becoming urbanised ($prob_{ri}$) is positively influenced; the further away from the Randstad, the lower $prob_{ri}$ is. The same goes for the distance to the nearest built-up area, although this effect is not significant. Further, when a transaction takes place in an area that is predominantly built-up, urban green or greenhouse horticulture, the transition probability is influenced positively. The 'green variables' tell us that the probability of a transition to red land use is lower when land is leased and/or when parcels are in an area where DLG has land development plans. Also, parcels lying in the provincial ecological main structure have a lower probability of becoming developed for urban land use than parcels not lying in this protective nature policy zone. Next, the influence of a buffer zone on $prob_{ri}$ is negative. Finally, being located in a Belvedere policy zone also influences the probability negatively.

9.5 Application of Model Results

9.5.1 Calculation of a Transition Probability Map

The model results can be used to calculate a transition probability map for all parcels – sold and not sold – in the entire Province of Noord-Holland (see Step 6 in Section 9.2). The map legend in Fig. 9.10 shows that the range of probability values is concentrated around the range [0:1]: from –0.46 to a maximum of 2.5, to be more precise, with the mean value at 0.53. There is no substantial difference between the mean values of this transition probability map and the computed transition probabilities for the observed transactions (0.40). An explanation for the probability values less than zero could be the presence of unobserved variables and, for instance, the fact that we are not able to include the variable 'leased land' in this final step as we do not know for the entire province which parcels are leased and those which are not. Another possible explanation is that our model does not account for the large impact of uncertainty resulting from land-use change over time: a buyer buys a parcel now and then has to wait for, for example, 10 years before a decision is taken concerning whether or not the land use of that parcel is allowed to change from rural to urban. So, in addition to uncertainty about whether the change will occur, the fact that the buyer determines his offer using a discounted rate also lowers his bid price and, thus, the transition probability values in our model.

To gain an insight into how the factors uncertainty and time play a role in our current model, consider the following: suppose a buyer has to decide whether or not to buy a parcel now, while knowing that after 10 years a decision will be taken whether or not the rural parcel may be developed. In our model, we assume for the sake of convenience that the developed or undeveloped parcel prices are offered at the moment just before this decision is taken. Suppose these prices are equal to the average transaction prices per hectare for the two subsamples: P_{red} = 198,300 euros/ha and P_{green} = 5,500 euros/ha. Now assume that the buyer estimates

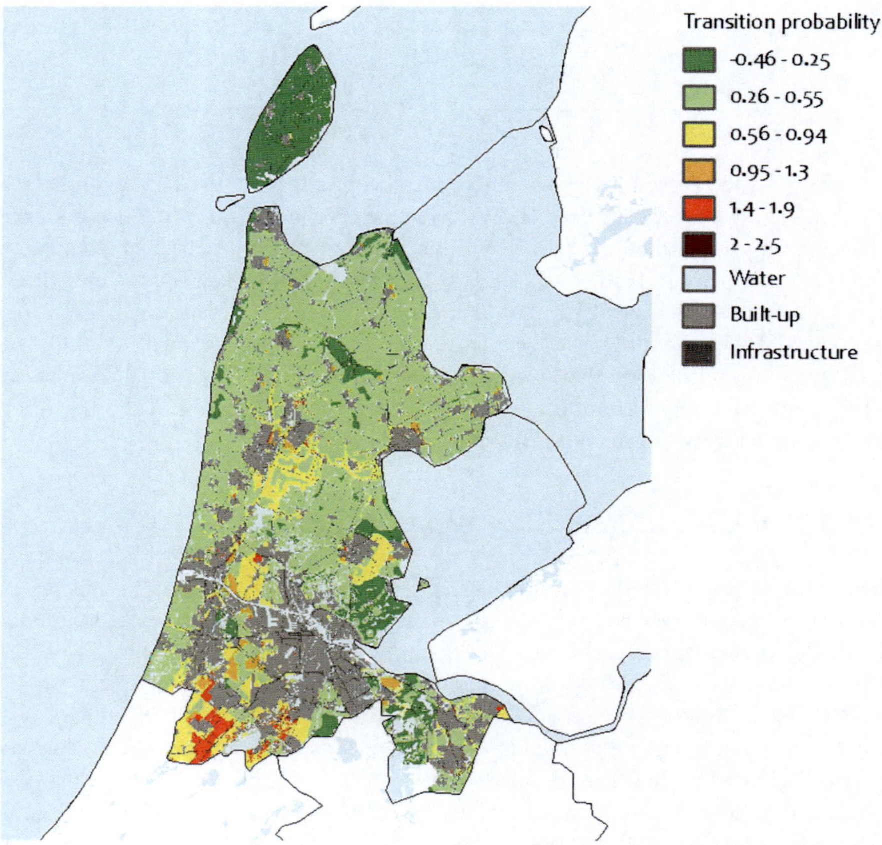

Fig. 9.10 Rural-urban transition probabilities in the Province of Noord-Holland (Source: Dekkers, 2010)

that he will have a 50% chance of being allowed to develop the parcel. Taking only this uncertainty into account, his bid price will be $0.50 \times (P_{red} + P_{green}) =$ 101,900 euros/ha. Suppose we now modify the model to allow for the decision whether or not he is allowed to develop the parcel is taken not now but rather in, say, 10 years time. In this case the buyer will incorporate an annual interest rate of, for example, 7% on his investment, to compute his bid price:

$$P_{green} = 385 + \frac{385}{1+0.07} + \frac{385}{(1+0.07^2)} + \ldots + \frac{385}{(1+0.07)^n} = 5,500 \quad (9.7)$$

where 385 is the annual revenues in euros per hectare ($0.07 \times 5{,}500$) related to the use of this land for green purposes.

$$\begin{aligned} P_{red} = 385 + \frac{385}{1+0.07} + \frac{385}{(1+0.07)^2} + \ldots + \frac{385}{(1+0.07)^9} + \\ \frac{13,881}{(1+0.07)^{10}} + \frac{13,881}{(1+0.07)^{11}} + \ldots + \frac{13,881}{(1+0.07)^n} = 103,500 \end{aligned} \quad (9.8)$$

where 385 is the annual revenues in euros per hectare for the first 10 years in which the parcel is still green, and 13,881 are the annual revenues in euros per hectare from year ten onwards, when the parcel will be used for 'red' purposes. This situation leads to a bid price of $0.50 \times (P_{red} + P_{green}) = 54{,}500$ euros/ha. The result is, thus, a considerably lower bid price than if this delay in development is not taken into account. If this lower bid price were to be used as input in our model it would lead to a transition probability that is considerably lower than 0.50. Thus the transition probabilities we measure reflect a combination of possible delay of the transition and the probability of such a transition in a narrow sense.

9.5.2 Relevance of Results for Land Use Scanner

The understanding of how the land market works and the occurrence of current land use, especially in the urban-rural transition zone, can help improve our ability to model future land-use change. The quantification of land-use change with the use of land-use models is very important for evaluating the effects of spatial policy (MNP, 2004; Borsboom-van Beurden et al., 2005). There are plenty of land-use models available for simulating land-use change. Most of them only simulate urban or rural land-use. However, because of an increasing overlap of policies in the Netherlands and the importance of rural-urban transitions in land-use change, it is desirable to have models that are suitable for integrated scenario analyses. For the Netherlands, several of these integrated models are available, one of them being Land Use Scanner, which is described at length in this book. This economics-oriented probabilistic model uses a logit-function in an iterative process to simulate demand for, and supply of, land (Hilferink & Rietveld, 1999). The way current and future developments, spatial policies and spatial pressure on land are modelled is subject to scientific debate. Over the past 2 years, ample attention has been paid to calibration and validation issues in the form of autologistic regression analyses, for example in the study by Loonen and Koomen (2009), which focuses on the probability of land-use change based on proximity to other land-use types. Further improvements can be made with regard to the sensitivity analysis of the results to the scale or resolution of modelling. A more fundamental improvement can be the restructuring, or enhancement, of the theoretical-economics foundations of the model, i.e. using information on land prices and land market processes in a more direct way in the model. Because of the big difference between processes in the land market and factors that affect land-use change, strengthening this theoretical-economics link is quite a challenge. However, we think that economics-based land-market models can provide the necessary theoretical foundations for these models. Consider the bid-rent theory and how, based on Alonso's urban land-market model (Alonso, 1964), McFadden (1978) developed an empirically more practical *stochastic* maximisation model that lies at the heart of Land Use Scanner.

Based upon our work, we propose three improvements to the way Land Use Scanner is currently used:

1. Land Use Scanner's algorithm computes probabilities for all land-use types in all cells separately and compares them as a kind of utility, where the probabilities/utilities are static. By using the results of multinomial logistic regression analysis (documented in Loonen & Koomen, 2009), we can improve the relative scaling of the different suitability factors for each land use's suitability map. For now, we have used only a limited set of factors in this analysis, which points to additional factors that help explain the transition to urban types of land use. This means that the results of the analysis can be used to select what other factors to include in the suitability maps for the multinomial logistic regression analysis. For example, in addition to location factors that are already included, such as the National Ecological Network and various accessibility measures, other factors that could be included are: distance to the Randstad, the presence of the land-use types 'urban green' and 'greenhouse horticulture', land development projects of DLG, Belvedere/UNESCO zones and/or Buffer Zones.
2. Since our analysis is based on actual land transactions in *real* euros, the results can also be used to rescale the suitability values for the different land uses in Land Use Scanner in *real* euros/m^2. Currently, the range of suitability values for the various types of land use are still based on expert judgement and vary from 5 euro/m^2 for nature up to 35 euro/m^2 for residential land. Further, being based on real euros also means that our transition probability map can be used as an initial bid-price map for urban land use (i.e. undeveloped land).
3. Finally, on a more fundamental level, we can strengthen the theoretical-economics link by examining processes in the land market using interview techniques and descriptive analysis (in line with the complementary role of institutional analysis mentioned in the introduction) to include information that can be derived from these analyses of bid prices for various land uses. For instance, we can examine how profits per hectare for different agricultural uses are translated by actors into actual bid prices in euros/m^2 and what differences there are between the average bid prices for different land uses so as to improve their relative scaling. An advantage of this approach is that it can also be applied to land uses for which no historical data (for the multinomial logistic regression analysis) are available (e.g. for biofuel production). This line of research is clearly linked to the work described by Kuhlman and others in this book.

The effect of directly including information from our analyses can be examined by defining a model configuration with only two types of land use: urban or rural. We further include the spatial data layers that are used in our 'red' and 'green' hedonic pricing models in Land Use Scanner. We then directly insert our HPM model coefficients of the red and the green models in the red and the green land-use suitability maps, respectively. If this approach is used, these suitability maps represent the initial bid of land-use type *j* for a parcel in cell *c*, thus improving the comparison of suitabilities, both between land-use types and for cells used by a single land-use type (Buurman, 2003). An additional advantage is that the suitability values can be directly related to market transactions in terms of euros per square metre, whereas the values currently used, although expressed in euros per square metre, are not based on such transaction data. Of course, in this example we only estimate one red and one green hedonic pricing model. With 10 modelled land-use types, this would mean constructing 10 hedonic pricing models.

On the basis of the above analysis, we conclude that there is scope to implement this approach in the current version of Land Use Scanner. This could be based on market transaction data, initially for, for example, only two aggregate land-use types. This would allow comparison of the original probabilistic algorithm of Land Use Scanner with the novel probabilistic approach we propose here. More specifically, it would allow us to examine what is the best way to transfer the knowledge derived from our approach into Land Use Scanner's allocation model: through replacing the model algorithm with our probabilistic approach; through using the probability maps derived as initial bid-price maps; or through a translation of the relative scaling of suitability maps into real euros. Finally, working with more realistic monetary values in Land Use Scanner would potentially shed more light on the derivation and actual values of the shadow prices that are generated in the iterative modelling process.

9.6 Concluding Remarks

In this chapter we have described our efforts to model the possible occurrence of spillover effects between segmented markets in the urban fringe. These sub-markets are created by spatial planning policies that restrict urban land development.

Earlier attempts to model the Dutch land market using hedonic pricing techniques only tried to explain what (spatial and non-spatial) factors contributed to land prices and/or tried to explain price differences between rural parcels with either an agricultural or urban designation. As described, we have taken these approaches one step further by developing a model that uses the hedonic pricing method in combination with a linear probability model. The resulting model uses the factors that explain differences in land prices to analyse the probability of rural land becoming urban land.

The results show that our approach satisfactorily answers a number of technical-methodological issues that occur in the models that originally gave us the idea for our approach: by not directly including different buyers in the analysis, we avoid the problem of assuming that all characteristics are equally important for all buyers. Furthermore, although the level of explained variance is not that high, when we try to explain what factors influence the probability of a rural parcel becoming urbanised, the factors in the Ordinary Least Squares estimation have plausible signs, are in general significant and explain their respective influence on rural-urban transition probabilities well.

Several recommendations for improving our model can be made. First, we expect that its explanatory power can be improved by, for instance, adding more explanatory variables, such as distance to railway stations, soil quality data and the share of a transaction that is located within areas defined by the European Bird and Habitat Directive. Second, the New Map of the Netherlands variables that describe where land-use change will occur should be replaced with variables that include information on the driving forces behind the land-use changes that occur. This could be done through literature study (e.g. De Nijs, De Niet & Crommentuijn, 2004; Verburg et al., 2004). Third, and perhaps most importantly, a new version of the model should include the timing of a transition, that is, how long does a buyer probably have to wait before the decision is taken whether or not the expected transition will take place? Procedures for changing the designation of a parcel can take many years, depending on, among other things, whether or not stakeholders submit formal protests against such a change. Explicitly including the factor time in the model is empirically relevant.

From an economics perspective, our analysis points to an important consequence of current spatial policy: the price gap at the urban fringe indicates that rural land just outside the urban fringe (observed from within a city) is used very inefficiently – alternative uses (i.e. urban uses) would yield higher returns on investment. Figure 9.11, in which the red and the green prices per m^2 of a parcel at the urban fringe are shown to differ by a factor of almost 24, illustrates just how large this price gap actually is.

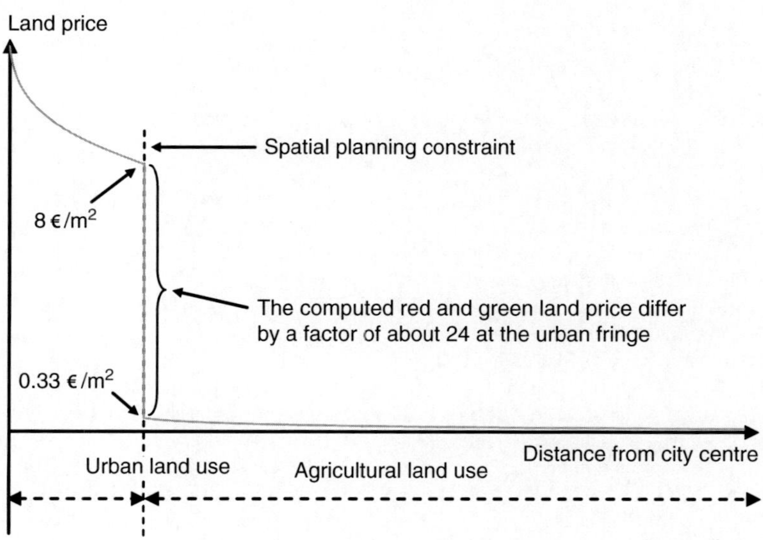

Fig. 9.11 Relation between red and green land prices at the urban fringe in the case of a spatial planning constraint (Source: Dekkers, 2010)

Acknowledgements We would like to thank the BSIK programmes 'Ruimte voor Geo-Informatie' (www.rgi.nl) and Habiforum (www.habiforum.nl) for partially funding this research. We extend, furthermore, our thanks to the Government Service for Land and Water Management of the Netherlands (DLG), a department of the Netherlands Ministry for Agriculture, Nature and Food Quality (LNV), for supplying the land transaction data. Thanks are also due to Jan Luijt and Tom Kuhlman of the Agricultural Economics Research Institute (LEI) of Wageningen University and Research Centre for discussions about land sub-markets, modelling and data issues. And finally, we thank Gerard Kooman, Edwin Bleijinga, Rik Heskes and Ton van Bart of the Province of Noord-Holland for discussions about the land market in their province.

References

Alonso, W. A. (1964). *Location and land use: Toward a general theory of land rent*. Cambridge: Harvard University Press.

Borsboom-van Beurden, J. A. M., Boersma, W. T., Bouwman, A. A., Crommentuijn, L. E. M., Dekkers, J. E. C., & Koomen, E. (2005). *Ruimtelijke Beelden – Visualisatie van een veranderd Nederland in 2030*. Bilthoven: Milieu- en Natuurplanbureau.

Buurman, J. J. G. (2003). *Rural land markets: A spatial explanatory model*. PhD Dissertation, Vrije Universiteit, Amsterdam.

CBS (2000). *Statline*, Centraal Bureau voor de Statistiek. www.statline.nl. Retrieved 15 June 2005.

Dekkers, J. E. C. (2010). *Externalities, land use planning and urban expansion*. PhD Dissertation, Vrije Universiteit, Amsterdam.

Dekkers, J. E. C., Rietveld, P., Van den Brink, A., & Scholten, H. J. (2004). *Exploring the land market in the province of Noord-Holland using a spatial regression model*. Paper for the 44th ERSA-congress, Porto, Portugal, August 25–29.

De Nijs, T. C. M., De Niet, R., & Crommentuijn, L. E. M. (2004). Constructing land-use maps of The Netherlands in 2030. *Journal of Environmental Management, 72*(1), 35–42.

Gordon, R. J. (1990). *The measurement of durable goods prices*. Chicago, IL: University of Chicago Press.

Griliches, Z. (Ed.). (1971). *Price indexes and quality change*. Cambridge, MA: Harvard University Press.

Hilferink, M., & Rietveld, P. (1999). Land use scanner: An integrated GIS based model for long term projections of land use in urban and rural areas. *Journal of Geographical Systems, 1*(2), 155–177.

Koomen, E., Dekkers, J. E. C., & Van Dijk, T. (2008). Open space preservation in The Netherlands: Planning, practice and prospects. *Land Use Policy, 25*(3), 361–377.

Loonen, W., & Koomen, E. (2009). *Calibration and validation of the land use scanner allocation algorithms*. PBL Report No. 550026002. Bilthoven: PBL.

Luijt, J., Kuhlman, J. W., & Pilkes, J. (2003). A*grarische grondprijzen onder stedelijke druk – Stedelijke optiewaarde en agrarische gebruikswaarde afhankelijk van ligging*. NPB Werkdocument 2003/15. Den Haag: LEI.

McFadden, D. (1978). Modelling the choice of residential location. In *Spatial Interaction Theory and Planning Models*, In A. Karlqvist, L. Lundqvist, F. Snickars & J. Weibull (Eds.), (pp. 75–96). Amsterdam: North Holland Publ.

MNP (2004). *Milieu- en natuureffecten Nota Ruimte*. Bilthoven: RIVM.

OCW, LNV, VROM, & V&W (1999). *Nota Belvedere*, Ministerie van Onderwijs, Cultuur en Wetenschap, Ministerie van Landbouw, Natuurbeheer en Visserij, Ministerie van Volkshuisvesting, Ruimtelijke Ordening en Milieu and Ministerie van Verkeer en Waterstaat, SdU, Den Haag.

RNP (1958). *De ontwikkeling van het Westen des lands*. Werkcommissie Westen des Lands van de Rijksdienst voor het Nationale Plan (RNP), Staatsdrukkerij, The Hague.

Rosen, S. (1974). Hedonic prices and implicit markets: Product differentiation in pure competition. *Journal of Political Economy, 82*(1), 34–55.

Segeren A. (2007). *De grondmarkt voor woningbouwlocaties. Belangen en strategieën van grondeigenaren*. The Hague/Rotterdam: Ruimtelijk Planbureau/NAi Uitgevers.

Segeren, A., Needham, B., & Groen, J. (2005). *De markt doorgrond. Een institutionele analyse van de grondmarkten in Nederland*. The Hague/Rotterdam: Ruimtelijk Planbureau/NAi Uitgevers.

Van Rij, E., Dekkers, J. E. C., & Koomen, E. (2008). Analysing the success of open space preservation in The Netherlands: The midden-Delfland case. *Tijdschrift voor Economische en Sociale Geografie, 99*(1), 115–124.

Verburg, P. H., Ritsema van Eck, J. R., De Nijs, T. C. M., Dijst, M. J., & Schot, P. (2004). Determinants of land-use change patterns in the Netherlands. *Environment and Planning B: Planning and Design, 31*, 125–150.

Chapter 10
A Policy Perspective of the Development of Dutch Land-Use Models

Marianne Kuijpers-Linde

10.1 Introduction

Preceding chapters show how land-use models have been used successfully in recent years to support decision-making processes for the formulation of new spatial policy strategies. In several of those chapters (see Chapters 4, 6 and 8), studies are described that explore the potential spatial patterns that may result from different policy alternatives related to, for example, climate change, the stimulation of bioenergy production and various strategies of urbanisation . This chapter focuses on future policy-related research questions that need to be addressed by the LUMOS toolbox. It describes, therefore, the LUMOS research agenda from a policy perspective.

To be able to define future policy-related land-use modelling questions, first of all insight is needed into current policy themes and the spatial planning relationships that are now being addressed in policy discussions. For example, the theme of depopulation is now receiving more and more attention. Until 2000, the expansion of urban areas was modelled and depopulation was seen as a liveability issue in peripheral rural areas, which was not included in spatial models. No explicit calculation rules were included in recent applications of Land Use Scanner for the depopulation of urban areas. Recently, this issue has also been linked to other issues, for example to urban renewal: is a concentration policy in urban areas the correct spatial planning strategy? A question that is now being asked as part of the depopulation discussion is: 'what are the advantages and disadvantages of building in an existing urban area to meet a need for expansion?' To answer such a question, planners must not only model an expansion of the urban area, but also changes in the urban area. This means that the spatial characteristics of changes in the housing market must be included much more explicitly in spatial models.

Second, to determine the modelling questions for the future, insight is needed into how policy is used to guide development. The points of leverage for policy

M. Kuijpers-Linde (✉)
TNO Urban Development, PO Box 49, 2600 AA Delft, The Netherlands
e-mail: marianne.kuijpers@tno.nl

E. Koomen, J. Borsboom-van Beurden (eds.), *Land-Use Modelling in Planning Practice*, The GeoJournal Library 101, DOI 10.1007/978-94-007-1822-7_10,

in the social and spatial system must be clearly distinguished in order to make a policy-relevant schematisation of reality – i.e. a model. The way in which spatial planning guides development changes over time and depends on the role that government plays in society. The Dutch philosophy of governance in land-use planning has, for example, changed drastically over the past few decades. Strategy embracing a centralised, development-control policy has been replaced by a one embracing a more localised, development-oriented policy. The different roles played by government can be illustrated with the policy themes of sustainable energy and environmental protection. To stimulate sustainable energy, government chooses primarily financial measures, whereas in environmental policy, government influences society primarily through the development of legislation and regulations. This implies that financial characteristics must be modelled when sustainable energy policy is the topic of research. However, if environmental policy is concerned then environmental standards based on spatial relationships between various types of land use should be included in the land-use model, for example as maps describing restrictions on various types of development.

From an inventory of current spatial planning themes and an analysis of the new role and instruments of government, it is possible to define the requirements the LUMOS toolbox must meet. To do so, one needs to look at the possibility of deriving meaningful indicators from land-use simulation results. Indicators are quantitative evaluation measures that summarise results, preferably in relation to specific policy themes (Bubeck & Koomen, 2008). Policy variables (points of leverage for policy in the spatial and societal system) and indicators are selected based on knowledge about topical policy themes and policy instruments. The flood-risk assessment that was performed using Land Use Scanner (Van der Hoeven, Aerts, Van der Klis & Koomen, 2008) offers an example of a useful indicator. This indicator relates to the current theme of climate change policy in spatial planning. One of the major aims of this policy theme is to limit the risk of flooding (expressed as economic damage or potential casualties). Flood risk is, thus, an important evaluation criterion (indicator) for spatial plans: policy can focus on preventing building on flood-prone locations. This means that the model must take into account the relationship between urban development (a policy-related variable) and flood risk at specific locations. Together with the organisation of knowledge input into spatial planning, such indicators and policy variables form the basis of the research agenda for the LUMOS toolbox.

The relationships between the LUMOS toolbox and policy are illustrated in Fig. 10.1. The development of a policy-support model begins by surveying the aims and policy instruments to be found in the policy for which the model will be used. Researchers transpose the policy aims into indicators (model output) and the policy instruments into policy variables (model input).

The structure of this chapter is as follows. The main, current spatial planning themes are introduced in Section 10.2. Section 10.3 then describes new developments in the policy instruments that are used to support the planning objectives. Finally, Section 10.4 discusses how the LUMOS toolbox should be adapted to help prepare new policies for current planning themes and related new instruments. It, thus, presents a number of modelling requirements that offer

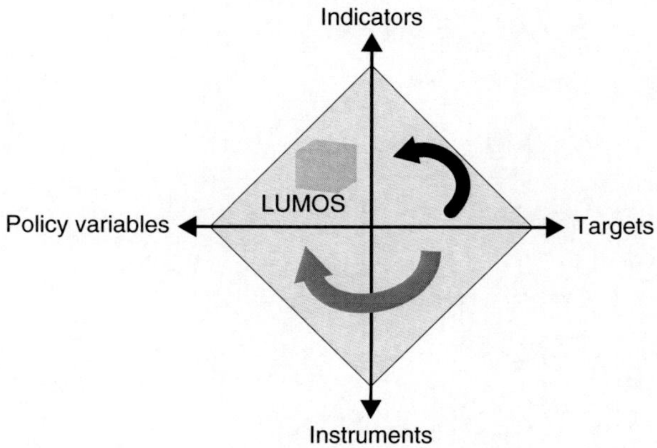

Fig. 10.1 Method for portraying relationships between policy variables, targets, instruments and indicators (source: Zondag & Borsboom-van Beurden, 2008)

directions for the further development of the LUMOS toolbox, which is discussed in the final chapter of this book.

10.2 Substantive Spatial Planning Themes

Themes in spatial planning typically follow developments in the economy, society and culture and are affected by dominant value orientation. Table 10.1 (Van der Cammen & De Klerk (2003) provides a clear overview of the history of land-use planning in the Netherlands, describing the most important spatial issues in relation to societal developments. The dominant land-use planning doctrines (values) described by Van der Cammen and De Klerk have been supplemented with topical policy dossiers of the departments involved. A striking feature of this summary is that since the 1980s a consistent choice has been made in the Netherlands for compact urbanisation and maintaining the urban-rural dichotomy (Faludi & Van der Valk, 1994; Koomen, Dekkers & Van Dijk, 2008).

Relationship Between Societal Issues and Spatial Planning

The issues on the spatial planning agenda have changed over the years and the overview presented in Table 10.1 clearly shows a succession of themes in the spatial planning dossier. For example, the expansion of the road network and the fanning out (or sprawl) of urban functions across the rural area in the 1970s, combined with the poor condition of inner cities, led to new attention for cities in the 1980s. In addition to the succession of themes on the agenda, interactions between societal developments and the way they are included in spatial planning are also apparent. Spatial issues are always linked to topical issues in society. The dossiers in the prevailing coalition agreement between the governing parties in the Netherlands for the period 2007–2010 are reflected in the spatial planning domain. For example, the financial crisis of 2009 has manifested itself in spatial projects as an investment issue.

Table 10.1 Chronological overview of the most important developments in spatial planning and related research

Period	Social order	Organisation of spatial planning	Dominant spatial concepts	Spatial analysis methods (examples)	Policy issues: Spatial developments
1930–1970	Development of the welfare state	Public programming, planning and investment Private implementation Centralisation	Functionalism Neighbourhood focus Dispersed urbanisation Urban region National road network Open cities and open countryside	Economic theory Spatial interaction Hierarchy of urban cores Quantitative policy support	Public housing: quantitative housing shortage Growing cities and new cities National motorway network Restructuring peripheral areas Ruralisation and redevelopment of rural area Delta works
1970–1985	Completion of the welfare state	Belief in the potential of planned change: public programming Land-use planning as a vehicle for welfare and prosperity policy Centralisation	Urban renewal Increasing scale and specialisation Urban regions (bundled deconcentration, new towns) Nature and landscape conservation	Integrated quantitative calculation models and traffic models Spatial distribution research (Lowry, Batty) Agricultural modelling Prognosis models (economy, traffic and housing)	Environmental issues (waste, acidification, water and soil contamination) Liveability (rural area) Economic issues: increasing scale and specialisation of urban regions Nature and landscape conservation Urban renewal
1985–2000	Post-modern social investment state	Loss of confidence in the potential of planned change Public-private cooperation Process planning Enforcement of land-using planning Increasing influence of market forces Decentralisation	Compact city and urban networks Restructuring Continuing integration of environmental and water policy Spatial quality and sustainable development	Specialised quantitative environmental and water models Decision support systems Sector-based reconnoitring of the future (environment, traffic, agriculture) Scenario research Separation of design and research Infrastructure effects Overview methodology (cost-benefit analysis)	Quality of living environment (urban area, quality of housing and employment) Energy supply Migration Continuing integration of environmental and water policy Agriculture and food safety Environmental issues (climate, air quality, noise). Innovation

Table 10.1 (continued)

Period	Social order	Organisation of spatial planning	Dominant spatial concepts	Spatial analysis methods (examples)	Policy issues: Spatial developments
2000–now	Risk society	Development planning Redefinition of tasks and responsibilities (new spatial planning act) Reorientation between market and government Emergence of regional governance as regional director Continuing internationalisation	Climate resilience (water assessment) Energy supply, layer approach and sustainability Compact city and urban networks: Delineating existing urban areas (consolidate and restructure) Emphasis on accessibility (nodes and multimodal networks) Separation of urban and green habitable areas Identity Risk-based approach Integrated regional development (financing)	Formalisation and statutory anchoring of relations between government and research Environmental impact assessment and water assessment National Advisers and Planning Offices Standardisation of models and theme portals (e.g. risk map for falling water-tables) Resource kit for uncertainty analysis Coordination of model chains and basic data registration	Security Energy supply Ageing and depopulation Restructuring existing urban area and business estates Climate resilience Cultural identity Landscape (space and peacefulness) Agriculture and food supply International competitive position and financial vulnerability Accessibility Sustainable development of Randstad
Future	Sustainable society	Selective national governance by means of core departments, spatial investment agenda and selection of themes and regions Implementation through regional governance External integration of culture in spatial planning	More attention for spatial quality expressed in guidance of spatial development and more attention for management and control Internalisation of sustainability in socio-economic system: sustainable production chains (cradle to cradle) Sustainability issue linked with lifestyles	More attention for design workshops Pictorial quality atlases Regional passports Sustainability impact assessments and climate guide Simplification of advisory and monitoring structure Development of regional models Focus on coherence between issues	Sustainable development (e.g. sustainable cities with residential districts) Sustainable production (e.g. urban agriculture, district heating and room for industrial activities) Energy transitions Climate adaptation Regional population decline Vacant office space Economic stagnation requires new investment strategies

Source: Adapted from Van der Cammen and De Klerk (2003)

Relationship Between Sectoral Policy and Spatial Planning

Policy developments in sectors other than spatial planning determine the topics on the spatial planning agenda, the underlying governance philosophy of that agenda and the set of instruments that are chosen. A good example of this is the external integration of environmental policy that has been implemented since the 1990s. The formal statutory link between land-use planning and various environmental dossiers (such as air quality and noise) has resulted in these issues remaining high on the policy agenda, despite major improvements in environmental quality. Another example is the strategic spatial vision for the Randstad (VROM, 2008), which is primarily linked to climate policy and traffic and transport policy.

Role of Public and Private Actors

Besides these interactions between various policy fields, the policy agenda in spatial development is determined by shifts in the balance between private and public actors and the distribution of tasks and competencies between tiers of government. Currently, the division of tasks between public and private parties is becoming less clear. Through their land policy, some municipalities are starting to act as entrepreneurs, while, in contrast, some private parties consider collective values in their investments. This implies that land-use models should pay attention to the changing agreements between public and private parties. In addition, collective values may receive an economic value in modelling applications.

Role of National Government, Provinces and Municipalities

Regarding the responsibility of the national government, provinces and municipalities, the National Spatial Strategy is very explicit: 'local if possible, central if necessary'. Based on current developments in many policy dossiers, it can be expected that the role of government in spatial planning will once again increase (due to the failure of market forces and the need for financial investment in collective values). This applies not only for the national tier, but also the regional tier (provinces and water boards). By means of a regional approach, a transition is taking place from diverse sectoral policy within the European and national restrictions to an integrated spatial planning strategy. At the national level, issues concerning security, climate and energy, water, nature, agriculture and landscape will require considerable centralised intervention.

To support further development of the LUMOS toolbox, publications of PBL's departments were surveyed and interviews were conducted (Kuijpers-Linde & Koomen, 2009). This resulted in a list of important and topical themes in the *spatial planning domain* (Table 10.2). In the future, spatial models will be used to support policy that will be developed for these themes. This application of spatial models partly relates to questions about future developments (exploring the future) and partly to questions about potential options for guiding spatial development (ex-ante policy assessments).

In the future, spatial planning will focus on the compatible and conflicting aspects of these issues. The LUMOS toolbox could be used in policy development if potential spatial developments are included in relevant model relationships, as will be illustrated in the following sections.

Table 10.2 Most important policy themes in the spatial planning domain

1. *Coping with climate change and energy supply.* Climate change requires strong interventions focusing on mitigation and adaptation, mainly related to water management. What measures are necessary? And when should these measures be implemented? Mitigation measures are clearly linked to the energy theme
2. *Integrating sustainability in spatial planning processes.* What does this mean and how should this be implemented?
3. *Accommodating the need for growth* in the Randstad *and managing stabilisation and depopulation* elsewhere. How can this trend be given shape in the new, general administrative orders and other instruments (investment plans, programmes and projects)?
4. *Intensifying the use of existing built-up areas, improving the accessibility of cities and strengthening main ports.* This issue relates to the demand for urban space and the positioning of the Netherlands in European and international networks. What space is available in rural areas and what does this mean for the liveability in cities and the countryside?
5. *Stimulating the agricultural economy.* How can agriculture be strengthened while taking into account new European policy and possibilities for using agriculture to conserve landscape qualities?
6. *Basing nature conservation on dynamic biodiversity targets.* In the context of climate change it may be necessary to adjust nature conservation policies to better fit the changing conditions. How can this be put in practice?
7. *Strengthening regional identity* (landscape, culture-historical values). This issue is closely related to the Ministry of Housing, Spatial Planning and the Environment's theme of *emphasising regionally oriented and differentiated policy.* Increasingly, planning issues require an integrated regional approach. For example, improving the spatial quality of the rural–urban fringe and strengthening agriculture as a carrier of rural qualities. Who will steer these processes and which division of tasks among different governmental layers is needed?

10.3 Policy Instruments for Dutch Spatial Planning

In the spatial planning domain, there has not only been a shift in substantive focus, but also in the way in which government tries to achieve these aims. The development of policy instruments has been strongly influenced by changes in values and standards. Government is no longer seen as the primary force that guides spatial development, but more as one of the actors who works together to influence spatial development in the Netherlands. The role of the various tiers of government and the relationship between government and society have been set down in the Spatial Planning Act (WRO). This legislation went into force in January 2008, while the policy changes resulting from it are being implemented in phases. Tasks and competencies have been more clearly defined and ambitions and strategy have been separated. The various tiers of government lay down their indicative regional plans in their spatial strategy, as well as establishing the legal framework for zoning plans in their general administrative orders (national government) or regulations (provinces and municipalities). In this new Spatial Planning Act the procedures have been significantly simplified. The new instruments it provides not only make it possible to act more quickly, but also allow the complex relationships between various policy tasks – based on the interests of the various parties – to be accounted for in policy. This is illustrated by the fact that the majority of the topics in the general administrative orders arise from sectoral policy.

A distinction can be made between the various roles played by government:

- organising;
- developing;
- researching;
- agenda setting.

Depending on the role taken by government, land-use models can be deployed in different ways, as is briefly explained below.

Organising Role

The organising role of government is specified in the general administrative order on spatial planning and in the regulations. Besides these spatial planning instruments, sectoral legislation also plays an important role. For example, the restrictions for spatial development are established in environmental legislation.

The Dutch government has taken a strong position on the protection of collective values. Examples include protection of valuable landscapes, protection of biodiversity and protection of the population against flooding and environmental risks. This protection is provided by:

- designating special areas (National Landscapes, National Ecological Network, flood defences);
- setting limits on spatial development, for example by defining the contours of urbanisation;
- guiding development by means of relative locations (spatial decision ladders, node policy).

In any case, the modelling instruments must take into account these procedures, which are set down in general administrative orders and regulations. This means that the calculation rules in land-use models can now differ more between regions: after all, each province emphasises different aspects when assessing spatial issues. It is important to survey the applicable laws and regulations.

Developing Role

Regional developments are used to realise ambitions for specific areas. Important characteristics of regional development are:

- an integral approach to issues, based on a vision that is shared by all actors involved;
- Public–Private Partnership (PPP);
- complementary cooperation and subsidiarity with other processes: all parties do what they are good at doing, and overlap with other processes is limited;
- a business case results from the characteristics specified above that specifies not only the targets (indicative plan), but also the means (a balanced budget or

operational plan): the new Land Servicing Act is an important instrument in this process;
- surplus profit (the amount remaining after the agreed profit allocation to the parties involved) is a central theme: the surplus is to be invested in the collective values of the region (landscape, security, nature, etc.).

Land-use models are very suitable for regional development because they can break down the complex relationships between many aspects of the region into components. As a result, it is possible to improve understanding of a problem, and therefore the understanding of tasks as well. The argumentation in a discussion between the parties can be supported by land-use models. For example, the relative effect of making changes in water resources management for agriculture, landscape and urbanisation can be clearly illustrated by map images and indicators, whereby existing agreements can also be taken into account. All other things being equal, alternatives can be evaluated as well. Land-use models can be used to map the effects of specific choices on the interests of the parties involved (possibly expressed as costs and benefits). During this process, a distinction can be made between short-term and long-term time horizons.

However, such applications require the following:

- the existing modelling instruments must be suitable for application at a regional scale;
- the effect of land ownership must be included in the model (linked to the budget) and the costs and benefits must be independent variables;
- it must be possible to model costs; and
- it must be possible to specify the tasks for each land use-function in both quantitative and qualitative terms.

Researching and Agenda-Setting Roles

Spatial models can be used in exploratory studies to work out conceivable scenarios. One example of this is the use of Land Use Scanner in the 'Second Sustainability Outlook on the future of the Netherlands' study (MNP, 2007, see also Chapter 4). For each policy target, modelling instruments were used to literally map out storylines about future development. By comparing the map images, it became clear where possible spatial conflicts could occur. In this way, the details of a policy agenda can be assessed for consistency and completeness.

Another example of the use of models for research and agenda setting is offered by the design workshops that were held as part of the 'Vision of the Randstad 2040' study (VROM, 2008 and Chapter 8 this volume). To support this creative process, explorations from the 'Prosperity and Living Environment' scenario study (CPB, MNP and RPB, 2006) were converted into map images. These provided support for developing new spatial planning strategies. The results of explorations such as these can be used to improve the model schematics and parameters of the LUMOS toolbox.

Besides providing these rules and the corresponding instruments, government can influence spatial development other ways too. For example, government can influence land use by means of financing (investment programmes, subsidies etc.), and by employing soft instruments such as design competitions, design workshops, regional process managers and public information campaigns.

10.4 Policy-Based Land-Use Modelling Requirements

From the discussion of policy instruments in Section 10.3, it is clear that policy developments present many difficulties for the further development of the LUMOS toolbox. Modellers face new challenges, especially relating to some of the new themes listed in Table 10.2, such as water management, urban restructuring and intensification, and new developments in the agricultural sector. In addition, new roles of government and related new instruments should receive attention in future model development. The knowledge requirements related to land-use modelling of these issues are discussed in the rest of this section.

Water Management

In the Netherlands, water is an aspect that must be included in many policy dossiers. For example, the nature policy dossier includes water quality and falling water-tables; the agricultural policy dossier includes water quality, water shortage and damage by waterlogging and flooding; the economics policy dossier includes risk of flooding. By means of spatial planning, the government can intervene in the hydrological system and in water supply and demand. The current version of Land Use Scanner does not explicitly model the characteristics of either of these factors. As a result, the effect of spatial development on risk of flooding, water nuisance, water shortage and falling water-tables has not been mapped out properly. Including the hydrological system in land-use modelling is an option that can be explored.

Urban Restructuring and Intensification

According to many observers, the era of expansion locations for big cities and the development of large business estates is over. Attention has shifted to intensifying, improving and transforming existing urban areas. Even without large expansion locations and new business estates, planners must still think about the desired spatial structure of the urban area and its interaction with the surrounding rural area. Networks play an important role in this process. Creating smart links between networks and coordinating these links (nodes) is a second theme that is central to many policy issues. Policy leverage points include the accessibility of locations, the qualities of the residential environment and the qualities of businesses estates. The mutual relationship between network development and other spatial developments must be included in new land-use model applications, from which greater insight can be acquired about the role of effective policy on the accessibility of various locations.

Developments in the Agricultural Sector, Partly in Relation to Water, Nature and Landscape

Model schematics are useful for dealing with themes relating to rural areas for which urbanisation is an exogenous variable. The central focus is the demand for 'smart' combinations of agriculture with societal functions such as water management, nature management, cultural history and recreation such that agriculture continues to develop within the given economic conditions and new EU agreements. Important policy leverage points are the economic structure of agricultural areas, the cultural-historical values of the rural area, the recreation potential of specific areas, the qualitative and quantitative characteristics of the hydrological system, and the changes in these abiotic conditions for nature.

The research agenda for land-use modelling is determined not only by the substantive planning themes discussed above, but also by the characteristics of policy and the role of knowledge in policy. Points of attention include more emphasis on regional policy and dynamic modelling and a clear specification of the role of land-use models in formal planning assessment procedures. These issues are discussed below.

More Emphasis on Regional Policy and Dynamic Modelling

Many complex spatial planning tasks require regional specification of policy. To support choices in regional specifications, it is important to picture the effects of alternatives. A land-use model can help depict these effects if relevant spatial developments are modelled in detail, the factor of time is made explicit, and spatial interaction (such as agglomeration and competition effects) and actor behaviour are included in the model.

Specifying the Role of Land-Use Models in Formal Planning Assessment Procedures

During the development of policy, the application of land-use models in formal assessment procedures must also be taken into account. Instruments that will become increasingly important in the future are: strategic environmental assessments (SEAs and sustainability assessments); social cost-benefit analyses (including collective values such as biodiversity, security, health and cultural history); and business cases for spatial investments. The use of these process instruments will lead to a convergence in the assessment criteria that are being used in policy discussions. It is important that investments in the LUMOS toolbox anticipate this development.

10.5 Closing Remarks

New challenges in spatial planning require modifications to the available LUMOS toolbox. Policy requires further specification of applications such as those for the various themes described in this section. It is unlikely that all new policy questions can be answered by a single integrated model. Yet, a start has already been made on substantial improvements to Land Use Scanner to make it better equipped to answer (part of) the questions discussed above. For example, in the 'Sustainability outlook on the Netherlands' (MNP, 2007) an attempt was made to make components of the model more dynamic and link these to a transport model (see also Chapters 4 and 5). The model was, furthermore, linked to a flood-risk model (Van der Hoeven et al., 2008) and was used in several SEAs of Regional Strategic Visions (see Chapters 7 and 8). All these applications show that Land Use Scanner can be further improved in the context of specific policy-related applications. The greatest challenges for the near future are to expand this model based on these new requirements, to renew these and other components of the LUMOS toolbox and, at the same time, continue to apply it in policy-related research. If expansion of the toolbox is primarily focused on the policy questions related to water management and urban intensification and restructuring, new elements to be incorporated are:

1. information on the location of activities and actors, and the objects in the physical environment they use for these activities, for example, land and real estate;
2. information on actor behaviour and its consequences for land use and the transport system, for example, location choices of companies, substitution in the housing market, and responses to transport policy measures;
3. information on the dynamics of use of geographical space;
4. a structural link between land use and the transport system;
5. improved linkage to hydrological models for the assessment of water management measures in relation to changes in land use.

During further development of the LUMOS toolbox, the organisation of the knowledge input for spatial planning must also be taken into account. In spatial planning, there is continuing formalisation of this knowledge input and ongoing professionalisation of knowledge management. For example, in the Netherlands the method for estimating the costs and benefits of spatial investments has been systematically described. When designing a spatially explicit land-use model, it is important to determine which role the method should play, which land-use categories must be distinguished and which spatial relationships are relevant.

References

Bubeck, P., & Koomen, E. (2008). *The use of quantitative evaluation measures in land-use change projections; An inventory of indicators available in the land use scanner*. Amsterdam: Spinlab Research Memorandum SL-07. Vrije Universiteit Amsterdam/SPINlab.

CPB, MNP and RPB (2006). *Welvaart en Leefomgeving. Een scenariostudie voor Nederland in 2040*. Den Haag: Centraal Planbureau, Milieu- en Natuurplanbureau en Ruimtelijk Planbureau.

Faludi, A., & Van der Valk, A. (1994). *Rule and order: Dutch planning doctrine in the twentieth century*. Dordrecht: Kluwer.

Koomen, E., Dekkers, J., & Van Dijk, T. (2008). Open space preservation in The Netherlands: Planning, practice and prospects. *Land Use Policy, 25*(3), 361–377.

Kuijpers-Linde, M., & Koomen, E. (2009). *Beleidsvragen en indicatoren voor een nieuw ruimtegebruiksmodel*. Amsterdam: Geodan Next.

MNP (2007). Nederland Later; Tweede Duurzaamheidsverkenning deel fysieke leefomgeving Nederland. MNP-publicatienr.500127001/2007. Milieu- en Natuurplanbureau, Bilthoven.

Van der Cammen, H., & De Klerk, L. (2003). *Ruimtelijke ordening van grachtengordel tot Vinex-wijk*. Utrecht: Het Spectrum.

Van der Hoeven, E., Aerts, J., Van der Klis, H., & Koomen, E. (2008). An integrated discussion support system for new Dutch flood risk management strategies. In S. Geertman & J. C. H. Stillwell (Eds.), *Planning support systems: Best practices and new methods*, 159–174. Berlin: Springer.

VROM (2008). *Nota Randstad 2040. Ministerie van Volkshuisvesting*. Den Haag: Ruimtelijke Ordening en Milieubeheer.

Zondag, B., & Borsboom-van Beurden, J. A. M. (2008). *Uitwerking project LUMOS 1.0 in 2008*. Interne notitie. Bilthoven: Planbureau voor de Leefomgeving.

Chapter 11
Developing a New, Market-Based Land-Use Model

Judith Borsboom-van Beurden and Barry Zondag

11.1 Introduction

This book describes the extensive experience that PBL Netherlands Environmental Assessment Agency and its partners have built up in more than a decade of using land-use models to support policy-making. The land-use models of PBL – Land Use Scanner and Environment Explorer – have in several large studies contributed substantially to the research findings and policy recommendations, making them a standard component of the analytical framework of outlooks where spatial dynamics are essential for sustainability and environmental quality in future. The performance of the models has been evaluated both by PBL and an audit committee of international experts; the findings of the committee are summarised in Chapter 2. The overall conclusion is that the models represent the state of the art for their current practice, but that to enlarge their potential contribution to policy questions and to address new policy challenges a substantial model redesign, or the development of a new model, is needed. Important challenges in doing so would be the inclusion of the behaviour of key actors in the model chain, the inclusion of essential feedback loops between different sectors and geographical scales, the structural inclusion of transport within the land-use model, and a better integration of water management and land use to be able to evaluate spatial adaptation strategies related to climate change. The insights from the different subject-specific research findings have been highlighted and processed to formulate several basic features to be included in the design of a new land-use modelling framework. In this final chapter, we describe the way forward for the development of such a land-use model. After an evaluation of achievements and drawbacks of the current model chain for the support of strategic policy questions, we discuss the ambitions and options for a new land-use model. Subsequently, the various activities carried out to arrive at the general specifications for the new model and the results achieved so far are highlighted.

J. Borsboom-van Beurden (✉)
TNO Behavioural and Societal Sciences, PO Box 49, 2600 AA Delft, The Netherlands
e-mail: judith.borsboom@tno.nl

E. Koomen, J. Borsboom-van Beurden (eds.), *Land-Use Modelling in Planning Practice*, The GeoJournal Library 101, DOI 10.1007/978-94-007-1822-7_11,

11.2 Evaluation of Land Use Scanner Applications

Since the first version of the Land Use Scanner was developed in 1997, the model has been applied in a wide range of policy assessment studies at PBL (see Borsboom-van Beurden et al., 2005; Koomen, Kuhlman, Groen & Bouwman, 2005; MNP, 2004a, 2004b; RIVM, 1997; RIVM & Stichting DLO, 2001, 2002; Van de Velde et al., 1997). The most recent applications of the Land Use Scanner model by PBL and its partners have been discussed in several of the previous chapters of this book, including the comprehensive 'The Netherlands in the future' study that is a part of the Second Sustainability Outlook (MNP, 2007; PBL, 2010).

The diverse applications of land-use models in policy assessment studies have been evaluated internally at PBL. In addition, an international audit committee reviewed both Land Use Scanner and Environment Explorer in 2007 (see Chapter 2). Both models have been judged by the international audit committee in 2007 as 'among the best in their field, and very useful for assessing policy impacts relating to the spatial distribution of land use at the national level' (See Chapter 2 by Harry Timmermans, Michael Batty, Helen Couclelis, and Michael Wegener, this volume). Nevertheless, the audit committee concluded at the same time that a substantial redesign of the model or the development of a new activity-based model would be required to overcome the drawbacks of the current models. In addition, from the overview of current policy questions in Chapter 10, it follows that reliable, more detailed model outcomes are needed at a sub-national, or regional, level, which cannot be fully provided at this time. This section summarises the main achievements and drawbacks of the current set up of the land-use modelling chain, including that present in Land Use Scanner.

A major achievement of the current Land Use Scanner model is that it incorporates an extensive knowledge base, developed by PBL and its partners, containing information about the drivers for and progress of land-use changes. This knowledge base is actually formalised in the data underlying the model, such as figures on future demand for land, allocation rules and about 400 geo-datasets representing the suitability of particular land uses. The knowledge base also comprises detailed information on the impact of current and potential policy measures, not only for spatial planning measures such as urban containment and designated development sites, but also for sector policies such as water management and nature development. As such, this knowledge base enables the model to chart the spatial consequences of current trends and the potential impact of policy measures in the form of '*what-if*' reasonings.

This is a truly integral knowledge base: all types of land use have been worked out in detail. Many models claiming to be integral are in fact only directed at urban functions. 'Integral' means then the integration of housing, employment and transport, but does not include more rural functions. One category of examples of this partial integration are the so-called LUTI models (Land Use and Transport Interaction models), as developed by Miller, Kriger, and Hunt (1998), Simmonds and Enchenique (1999), Zondag and Pieters (2001), and Wegener (2004). The category of 'Spatial Computed General Equilibrium' models, which focus on the regional economic developments from a systems perspective, as developed by Oosterhaven, Knaap, Ruigrok, and Tavasszy (2001) and Bröcker (2004), also exemplifies this partial integration. Other families of models are mainly directed at rural functions so that urban functions are somewhat neglected; the CLUE-S model (Verburg and Overmars, 2009) is one such model.

Another achievement of Land Use Scanner is that the spatial elaboration of diverse trends and policy measures using a land-use model has really contributed to the preparation of strategic policies by 'framing' possible futures. Different aspects can be highlighted here. It appears that the application of a land-use model often turns out to work as an integrating framework: it visualises the spatial consequences of various trends and policies which might have otherwise remained unknown. Land-use models show where contradictions and frictions may occur, and where more coherence between policies could be pursued. Especially when policies are developed separately in specific sectors without considering the impact on other sectors, such integration can be very valuable. This support of the interactive process of policy preparation and policy-making by 'raising awareness' has proven to be as important for policy makers as the actual model outcomes (Guhathakurta, 2003). Further, results of land-use modelling give insights into the pressure that is placed on weaker functions such as agricultural land, valuable landscapes, nature areas, and flood-prone areas. This pressure might be contrary to the spatial, environmental and ecological policies pursued for these areas. What is more, land-use models can show where different land uses might be combined or optimised. By mapping the best options under certain conditions ('*what-if*'), the possibilities for optimisation of spatial structure and land use become clearer. Both conflicting and optimising outcomes can set the agenda for strategic policy-making by signalling desired or undesired spatial changes.

The most comprehensive example of such a spatial elaboration of trends is the 'Second Sustainability Outlook' study (PBL, 2010; see also Chapters 4 and 5), which contributed to the preparation of the Randstad Vision 2040 of the Dutch Ministry of Housing, Spatial Planning and the Environment (VROM, 2008; see also Chapter 8). By investigating the demand for land in the congested Randstad conurbation for several decades to come, and to show which land-use changes could be expected and how land use could be optimised according to various sustainability goals, the development of a new spatial strategy for the Randstad area was supported. Another important contribution was an extensive investigation of adequate responses to the risks posed by climate change that was commissioned by the Dutch Cabinet. The land-use changes simulated in the 'Second Sustainability Outlook' study clearly showed the concentration of demographic and economic growth in the western part of the Netherlands, and the increasing potential damage in cases of flooding (Delta Commission, 2008).

These examples of the contributions of land-use model outcomes in key policy documents demonstrate that Land Use Scanner has been successfully applied in the preparation of strategic policies in the Netherlands; in many other countries land-use modelling has remained a predominantly academic activity with a strong methodological-technical focus. Despite these successes, internal evaluations and an international audit have brought a number of drawbacks to light (Borsboom & Zondag, 2009):

1. *Lack of information about actors, activities and spatial objects apart from their land consumption*
   In a number of studies, it appeared that not all environmental and spatial problems could be addressed adequately because the current model disregards the distribution of households and persons, businesses, employment, farms, etc., when it simulates land use in the future. Often this information is available at an earlier stage in the model chain, for example, in sector models at a sub-regional scale. However, only their translation into future claims for land is processed in Land Use Scanner, usually assigning it to coarser spatial units. As a result, various indicators that require information on the future spatial organisation of households and businesses, such as exposure to air pollution, congestion arising from transport measures and the potential casualties and economic damage resulting from flooding, cannot be properly assessed. Although a workaround was developed for the 'Second Sustainability Outlook' study that assigned data on population distribution after land used was allocated, this solution was not ideal. Besides, under the current set-up the preferences and behaviour of actors are simulated indirectly by means of suitability maps for a specific type of land use, instead of being directly modelled via the location choices of households or firms. A more direct simulation of the behaviour of key actors, such as households, firms or farms, is needed to reflect the variety in responses of different types of households, firms or farms, to changes in circumstances, such as the physical environment or higher energy costs.
2. *Lack of dynamics in land-use simulation*
   The static approach of Land Use Scanner, which follows on from equilibrium modelling, is not in line with the usually gradual, path-dependent, changes that occur in land use. What is more, equilibrium modelling is mainly motivated by economic theory: the existence of an equilibrium in practice has not yet been proven. It can be argued that such a general equilibrium does not exist at any moment in time, due to various time lags and differences in the pace of change for various parts of the spatial system. This has been demonstrated by Wegener (2004), who gave an overview of the different time spans of change in the nine elements of the urban/regional system. A dynamic approach, allowing for partial equilibriums in specific markets, is therefore to be preferred as it is much more capable of addressing the relatively small changes in land use and the path dependency of these changes (Simmonds, 1999; Zondag, 2007).
3. *Lack of consistency between sector models mutually and between sector models and the land-use model*
   As we have noted, sector-specific models providing input for land-use models were developed for a particular purpose: for example a housing market model for making forecasts of future housing needs, or an agro-economic model for analysing agricultural profitability. Therefore it is not surprising that the coherence and consistency of assumptions, spatial units, definitions and lines of

reasoning in all these models can be improved significantly (Dekkers & Koomen, 2006). A major point of improvement is the inclusion of the mutual relation between employment, housing and transport, as Timmermans (2007) identified for the Welfare, Prosperity and Quality of the Living Environment study (CPB, MNP & RPB, 2006). An additional problem to the limited coherence and consistency between models in the model chain is the unidirectional character of the model chain in which Land Use Scanner is applied, not allowing the modelling of feedback mechanisms which exist in reality. As a result, changes in the availability of land and restrictions for specific land-use types are not dealt with in the sector models. Essential feedback mechanisms are lacking, such as substitution in the housing market for other types of housing or other locations, in response to housing shortages in particular cities.

4. *Insufficient elaboration and quantification of indicators*
   Many indicators cannot process Land Use Scanner's output adequately. Causes are highly divergent spatial units for analysis, incompatible definitions and the lack of information on actors or activities needed for assessing exposure to risks.

Considering the drawbacks it was decided to fundamentally redesign the Land Use Scanner model.

Prerequisites for this new LUMOS model are:

1. *The model must simulate for the whole of the Netherlands*, so the model has to be geared to national and regional analyses and is not designed for local analyses;
2. *The model must be fully multi-sectoral*, meaning that all relevant land uses, urban as well as rural, have to be covered;
3. *The model must clearly distinguish trends from scenarios,* by extrapolation of current developments or well-supported assumptions for important variables such as housing need or economic growth. The model has to be fitted for both types of applications.
4. *The model must simulate simultaneously land use, actors/activities and the objects needed for these activities;* these layers have to be integrally addressed in the model design, while essential feedback, such as transport in relation to housing and employment have to be included.
5. *The model must be dynamic and simulate intervals of one or 2 years,* which shifts the emphasis from modelling of equilibriums to the modelling of changes in time, allowing path-dependency and temporary impacts – for example, the impact of policy measures between 2015 and 2025.
6. *The model must be developed by PBL jointly with preferred external partners.*

Work on the requirements of the new model has entailed the charting of strategic policy questions, identification of the most important spatial processes and drivers, exploration of various options for the model design, and improvement of the calculation of various indicators, especially those for flooding risks.

11.3 Requirements of the Model: Policy Questions and Indicators

At request of PBL, Geodan Next explored the policy questions that were expected to potentially need outcomes of integrated land-use modelling (Kuijpers-Linde & Koomen, 2009) in the coming 5 years. This investigation has been discussed in detail in Chapter 2. The relevant policy questions found, which are closely interrelated, can be summarised in three coherent blocks:

1. *Urban restructuring, transformation and liveability*
   a. What are the possibilities and limitations for densification, urban restructuring and urban containment?
   b. How can more sustainable cities be created, for example in regard to their space-efficiency and energy-efficiency?
   c. How to respond to the impact of demographic processes on urbanisation patterns, such as population decline and immigration, from the viewpoint of segregation and liveability?
   d. How to reduce congestion and environmental pressure in the cities?
2. *Water management and biodiversity*
   a. What are the risks resulting from climate change and what needs to be done to further mitigate for and adapt to climate change?
   b. What will be the impact of changes in the energy system on spatial organisation and land use, and what are the spatial consequences of alternative energy sources?
   c. How to maintain a water management system in which both the quality and quantity are rated good?
   d. How to prevent further losses of biodiversity, also taking into account the consequences of climate change, and how can biodiversity policy aims be made more flexible so that it is able to respond to climate change?
3. *Retaining the identity and aesthetic value of Dutch landscapes*
   a. How to prevent further deterioration in Dutch landscapes?
   b. How to reinforce the vitality of Dutch agriculture without harming nature and cultural landscapes?

From this overview, it can be concluded that since the middle of the 1990s there has been a shift in the policy questions being dealt with: from the accommodation of a growing and competing demand for land to a better management and utilisation of land already in use by a particular function. Besides, it appears that to be able to answer many of these policy questions, reliable model outcomes are needed at a regional or even sub-regional scale, in addition to the national level. These outcomes should also include more information than just land use, namely information on actors/activities and objects, to be able to produce answers for these policy questions (Borsboom & Zondag, 2009).

11.4 Requirements of the Model: Driving Forces of Land-Use Change

In order to create an overview of the main driving forces behind changes in spatial organisation and land use, a number of experts were invited to participate in working groups for the themes of housing, businesses, agriculture, water management and nature management. In addition, PBL experts carried out research to identify overarching themes such as transport, leisure and energy (De Niet and Van Middelkoop, 2009; Jacobs, 2010; Rijken, 2009). Topics addressed were the major driving forces of spatial changes, their spatial materialisation, changes in spatial distribution and intensity of land use, and international aspects such as the influence of European and global developments and policies on spatial organisation and land use. All working groups and experts concluded their investigations with a paper describing the main spatial dynamics for their sector and the most important drivers (see Atzema, Korteweg, Lambooy & Van Oort, 2009; Priemus & Hoekstra, 2009; Van Bruchem & Silvis, 2009; Groot et al., 2009). Zondag and Borsboom (2009) summarised their main findings, which are presented in Table 11.1.

This summary of drivers was used to identify the following requirements that the new land-use model should meet (see Zondag & Borsboom, 2009 for a complete overview):

- *A shift in focus from modelling quantity to modelling quality*. Such a shift includes more attention for the modelling of characteristics of existing spatial structures and the processes of transformation or renewal of these structures. Traditionally, land-use models at PBL focus on changing land-use types, such as urbanisation on agricultural land. More attributes of the existing spatial structure and built-up area will have to be included to be able to model its relevant qualitative aspects, for example land and property ownership, social composition and value of biodiversity.
- *Increased possibilities for simulation of regional diversification*. It is expected that in the coming decades the Netherlands will face patterns of regional growth and shrinkage in terms of population and employment. These processes will also have an impact on the functionality of areas, since they affect the potential for large-scale agriculture, nature development, leisure (day recreation and holiday homes), public transport and other services. Traditionally, land-use models focus on growth: the demand for land from various sectors is allocated to scarce land resources.
- *Integrated and simultaneous modelling of land use (for example agriculture), objects (for example farms) and actors (for example farmers)*. In particular, the inclusion of some characteristics of the main actors, allowing for some detail in the modelling of their responses to the driving forces of land use change,

Table 11.1 Overview of the main drivers of land-use change

Housing	• Demographic drivers remain important; a new phenomenon for the Netherlands is population shrinkage in more peripheral regions. Other issues are an increasing variety in household types, migration and the behaviour of migrants • Increasingly important are economic and financial drivers, such as household income and capital, and financial instruments of the government • On the supply side, the existing spatial patterns (e.g. houses, employment and infrastructure) and policies are dominant. Changes occur more and more within the existing housing stock • Relationship between supply and demand is becoming more and more complex (definition of a house, second homes)
Businesses	• Drivers and their impact differ strongly by sector (industry, transport and distribution, consumer or business services) – economic structure effects play role too in the demand for land • Besides traditional location factors (labour cost, transport cost and land prices), soft location factors are becoming more important (skills of employees, access to knowledge, residential amenities, image etc.) • Strongly interact with demographic developments in regions and transport facilities • In the real estate market there is a shift in focus on quantitative growth towards qualitative improvements, as the renewal and restructuring of existing sites and buildings. Other dominant trends are concentration and densification
Agriculture	• Conversion from agricultural land to urban land or nature areas strongly driven by developments exogenous to the agricultural sector • Changes in type of agricultural production (crop, cattle and horticulture) are mainly influenced by developments on world markets, technological developments and environmental restrictions • Multi-functional land use is being driven by changes in income support by European Common Agricultural Policy, stressing more and more nature and landscape management instead of further rationalisation and rise of production, but also by societal changes as the increasing importance of recreational activities in the countryside and a rising number of hobby farmers
Nature and water management	• Societal and economic drivers such as outdoor leisure activities or short stay vacations and housing at the water's edge influence quantity and quality of both water and nature • Climate change influences land use via its impact on physical drivers, such as intensification of rainfall and drought, rise of the sea level and salt intrusion, and via adaptation strategies (additional water storage and development restrictions) • Quantity and quality aspects of nature and water management interact strongly with developments in the agriculture sector, leading to scale increases and regulation of water levels and emissions • EU policies are becoming more and more important, e.g. EU water framework, Natura2000 (and the national ecological network) and the nitrate guideline

is needed. It is known that within the same class of actors (such as households, companies and farmers), people respond differently to the same drivers, depending on their characteristics. For example, a classification of households by income is needed to adequately model their responses to road pricing strategies. Another example is the classification of households by size, since a household with children is likely to prefer other types of housing and residential locations than a household without children. The future use of geographical space in terms of objects and land is highly influenced by relative shifts in the breakdown of categories, for example in population composition, socio-economic background, and economic structure. Changes in these breakdowns such as more or less households with children, more or less high income households, or more offices and less manufacturing, all have an impact on the use of geographical space. The reason for this is that variables explaining that use, such as accessibility or urban green, work out differently in the location choices of each category of actors.

- *More thorough modelling of the impact of climate change*. Climate change affects physical conditions in the Netherlands as a result of, for example, rising sea levels, periods of heavy rainfall and drought, and increasing seepage of salt water. As such, it affects the amount of land needed for nature, water management and agriculture, but also the way this land is utilised. In addition, flood risk for urbanised regions along coasts and rivers is likely to determine future land use. The water system plays a crucial role when calculating the impacts of climate change or addressing the effectiveness of adaptation strategies. Therefore the interactions between water and land use in the broadest sense should be included more fundamentally in any new land-use model. In the current model chain, however, mutual interactions between water and land use are virtually lacking. Although several maps are available that describe locations with a reduced or enhanced suitability for specific land-use types, e.g. groundwater protection areas that limit the suitability for new housing, water management is poorly represented. Surface water is exogenously planned for as a land-use category, and information on hydrological relationships is not included at all. For the new LUMOS model framework, it will be important to include such hydrological relationships to enable the establishment of upstream and downstream relationships. Insight in the hydrological relationships could be used to restrict upstream developments at certain locations to protect downstream functions. Or alternatively, this information could be used to select, for example, suitable locations for nature development.
- *Attention for the impact of energy systems*. Energy transition, driven by climate-change mitigation policies and high energy prices, is an important and sensitive scenario variable, affecting the land use of different sectors in diverse ways. Land can be used directly to produce energy (windmills, biomass, etc.) and to distribute it. The indirect impacts of energy-related changes, for example through mobility or housing costs, are potentially large because they affect the spatial organisation of production and consumption chains and the costs of construction.

11.5 Options

11.5.1 Exploring Options

One can conclude from Sections 11.3 and 11.4 that the set-up of the current model chain no longer suffices for answering current policy questions. When the LUMOS project started in 2008, the initial intention was to build, conform the advice of the international audit committee, a 'full-fledged integrated agent based simulation framework', which would in the long term replace Land Use Scanner. Such a framework could make use of state-of-the-art insights, methods and techniques in spatial modelling, such as micro-simulation, and could enable the simulation of mutual interaction between actors or processes as firm demography (Van Wissen, 2000). A micro-representation of activities in space would also allow feedback processes that take place at a local level to be simulated. An example of such a feedback process is the influence of traffic-based noise pollution on the attractiveness of residential locations.

A micro model often requires the creation of synthetic data to represent the individual level of people, households and firms. In addition, stochastic variation is introduced to simulate decisions at the micro level, which introduces the problem that multiple model runs based on the same assumptions and causal relations, result in different outcomes. This is not desirable when policy questions have to be answered: model outcomes usually have to repeatable in such cases. Similarly for disaggregate types of models, the coefficients for the explanatory variables in micro models often need to be estimated on a larger population in order to have enough observations. Further, because of the large data volumes and potentially long calculation times for a micro model, complex software will need to be developed to ensure efficient data management and acceptable run times.

The development of a completely new, actor-based integrated model, as preferred by the international audit committee in 2007, is no longer considered to be a realistic option given the time-span, risks and costs associated with it. Examples of such model developments as TLUMIP, ILUTE and ILUMASS (Salvini & Miller, 2005; Wagner & Wegener, 2007; Weidner, Donelly, Freedman, Abraham & Hunt, 2006) show that, in spite of the large effort in terms of labour and costs, these models are not yet fully operational (ILUTE, ILUMASS) or are still scarcely used for policy-making (TLUMIP). Therefore the applied methodologies and techniques cannot be considered as proven technology. The high expectations for the added value of actor-based techniques for the modelling of complex spatial systems have so far failed to materialise in actual policy-making (see, for example, Ligtenberg, 2006; Ettema, Floor & De Jong, 2009).

In the meantime, the existing PBL model infrastructure has grown much larger due to a merger between MNP and RPB (Netherlands Bureau for Spatial Analysis) in 2008, resulting in doubt about the approach to take. In particular, it raises the question as to whether LUMOS could also be developed by better utilisation of parts already developed within the PBL model infrastructure, such as the demographic model PEARL (RPB/CBS, 2005) and the employment model WEBER (De Graaff, Van Oort & Florax, 2009). Furthermore, PBL has become co-owner, together with the Ministry of Transport, Public Works and Water Management, of the TIGRIS XL land-use and transport interaction model.

Two options for building the LUMOS model framework along these lines were explored: building the framework around the LUTI model Tigris XL and expanding the Land Use Scanner model. By choosing Tigris XL as the base for further development of the LUMOS model framework, nearly all requirements of the new model as formulated in Section 11.4 can be fulfilled at lower costs, with less development time and at less risk, while the option to expand the Land Use Scanner model does not lead to the fundamental improvement of the model advised by the audit committee. Although this implies that actor behaviour is not represented at a micro-scale but in an aggregated form in Tigris XL, it still allows for the representation of heterogeneity in actors and their behaviour, for example by distinguishing various types of households exhibiting different responses. This set-up also allows for the representation of activities and objects at a detailed geographical scale. The structural coupling to the underlying transport model enables the assessment of the impact of policy measures such as road pricing and investment in public transport. In addition, the Tigris XL model is already dynamic because it produces results in time-steps of one year. Moreover, this option opens up possibilities for a better connection to sector models such as the demographic model PEARL, the employment model WEBER, and the housing model HOMERA. As a result, this option is well equipped to answer most policy questions concerning urban transformation and restructuring, e.g. possibilities for densification in built-up areas, accessibility benefits of urban containment, and the space-efficiency of urban sprawl. Borsboom and Zondag (2009) consider the policy questions relevant for LUMOS in more detail.

The Tigris XL option also brings with it the additional advantage that elaboration – expansion and deepening – of the model framework can be done gradually, by adding modules or meta-data that can be worked out in more detail at a later stage. Therefore the development of the new LUMOS model framework by elaboration of the Tigris XL model has been proposed as the preferred option for developing a new LUMOS model (internal note, Borsboom-van Beurden & Zondag, 2009).

11.5.2 Expanding Tigris XL

This option does not aim to build a very detailed 'state-of-the-art activity-based land-use model', but to build a 'state-of-the-practice market-based land-use model' (Miller et al., 1998). In the latter sort of model, the segmentation of the population or firms in disaggregated groups should be detailed enough to make it possible to simulate the main differences in behaviour. Examples of this are the different responses of high- and low-income car owners to the levying of tolls, or the different strategies of farmers towards the reform of EU Common Agricultural Policy. To develop this option, new model architecture is needed in which land use is treated dynamically and actor behaviour is simulated more explicitly and at a lower scale than that used so far. For many policy questions, it is precisely the intra-regional developments that are of crucial importance: examples are issues such as possibilities of urban densification, the impact of agriculture on biodiversity, and the consequences of measures for mitigation of climate change on activity patterns and transport flows. These developments require much more information on the housing market, labour market, transport system and rural functions than is available at the moment from Land Use Scanner.

The most obvious choice for realising the framework would be to expand the existing Tigris XL model, which already includes integral architecture for urban functions. The model simulates interactions between demography and housing, employment, transport and, to some extent, the availability of land. Other sectors such as agriculture, nature and water management first have to be added to the framework. The building of the LUMOS model framework can be organised using a modular architecture, which allows for a step-wise, gradual extension and improvement of the model's framework. To be able to simulate not only land use but also the distribution of actors/activities and objects, more knowledge from sector models, such as those on demography and housing, needs to be incorporated in the modules.

Then the match between supply and demand for land and objects can take place efficiently at the scale where it actually occurs, usually at a regional level, corresponding with the spatial unit applied in the sector models, or an aggregation of it. This integrated model framework should interact in time with an underlying GIS database of grid data containing information on suitabilities, which determine the availability of land for specific uses. In this way, factors defining the supply of land at grid-cell level, such as designated nature conservation areas or greenhouse concentrations, are translated into the higher scale, where supply and demand are matched. At the same time, changes in the demand for land, for example as a result of the approval of a plan for the construction of housing, are translated to this lower level of the individual grid cell (see Fig. 11.1). Such a set-up takes advantage of the fact that the sector models are in general strong in the modelling of the demand for objects and land, while the supply of objects and land is often

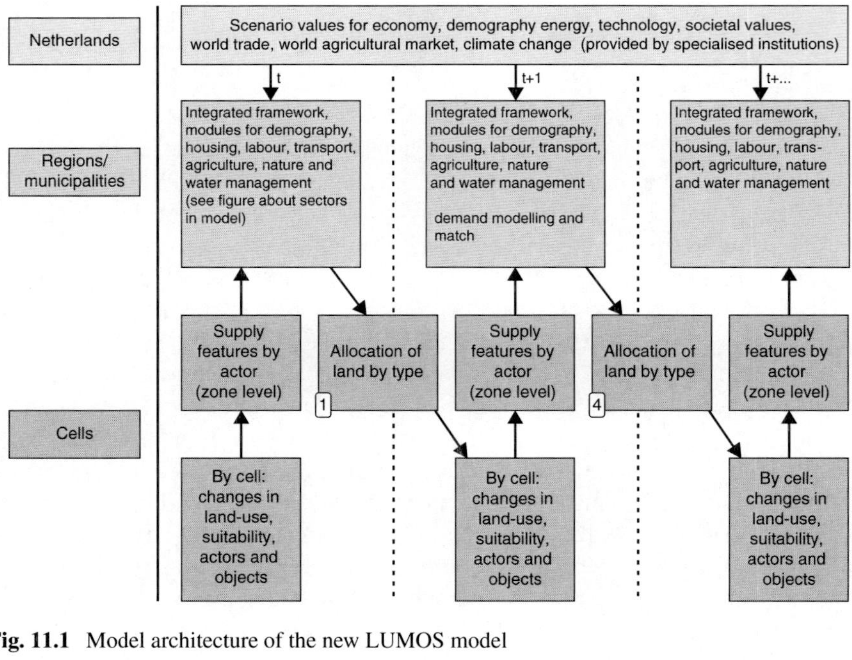

Fig. 11.1 Model architecture of the new LUMOS model

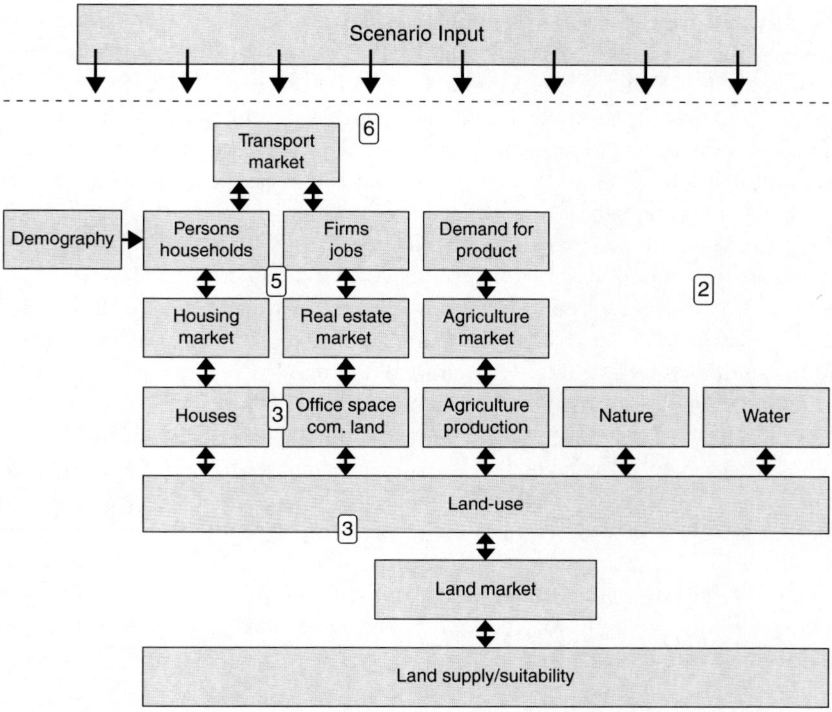

Fig. 11.2 Overview of the sectors incorporated in the new LUMOS model framework

poorly represented. Land Use Scanner currently contains highly-detailed spatial information on the suitability and availability of land, through GIS-data on soil, spatial planning, groundwater protection, noise pollution, etc. The modelling of supply in sector models can be enhanced by using this information in an aggregated form at the regional level.

To reduce investment costs and risks for the development of this framework, where possible, existing models and data will be used. Based on all these elements, a global design of the LUMOS model framework has been drafted and is presented in Figs. 11.1 and 11.2 although this description is clearly only a very rough sketch, which definitely needs to be worked out in more detail before the system can be built. Recently, databases and specifications of the demographic model PEARL were inserted in TIGRIS XL for the Spatial Outlook project. This could be a first building block for the framework.

11.6 Results So Far and the Road Ahead

Since 2008, the various research initiatives in the LUMOS project have roughly defined the contours of the new LUMOS model. The study by Geodan Next clarified the policy perspective (see Section 11.3 and Chapter 10) and the thematic expert groups charted the main drivers behind spatial changes (see Kuijpers-Linde & Koomen, 2009; Priemus & Hoekstra, 2009; Atzema et al., 2009; Van Bruchem & Silvis, 2009; De Niet & Van Middelkoop, 2009; Jacobs, 2010; Rijken, 2009; Groot et al., 2009). In addition, Ettema et al. (2009) explored the architectures of comparable systems and their application in practice. Their study confirmed that in a policy setting, such as that of PBL, developing a micro- or agent-based model is too risky in view of the length of time needed for development, the huge data demands, the long computation time and the increasing detail of the behavioural models. Their recommendation, therefore, was to tailor the level of detail (or breakdown in classes of agents) to the policy scenarios to be evaluated. Furthermore, the ICT aspects of the development of such a model or model framework have been explored by Hilferink and Grashoff (2009). They drafted a check-list of recommendations for the integration of various model components within a framework.

Now that the option to build the framework around the Tigris XL model has been chosen, measures to implement this decision in the next years have to be taken. The development of the new LUMOS model will take place in a stepwise fashion. Seven essential steps for realising a first version of the new model as presented in Figs. 11.1 and 11.2 are described below. The numbering of these developmental steps corresponds with the numbers in the figures, indicating which part of the framework is improved by that particular step. The following steps are foreseen in the short term (2010–2011):

1. *Preparation of the base year*
   First of all, to be able to make simulations using the model prototype, the required data on actors/activities, objects and land use have to be collected for a specified base year. GIS-data on land use and suitability from Land Use Scanner can be used at the grid cell level of 100 m × 100 m. The data on actors/activities and objects needed at this level must be derived from geographically very detailed datasets, such as business registrations (LISA), address data (CAN), and farm data (BIN). After being collected, the data will have to be harmonised with and made compatible to the land-use data and other data at the regional level. For example the total amount of land used by agriculture according to the Agricultural Census is not corresponding to the total amount of land used by agriculture according to the Land Use Statistics dataset from Statistics Netherlands.
2. *Addition of modules lacking in the framework*
   The current Tigris XL model contains modules for demography, housing, transport, employment, real estate and land. For carrying out truly integral analyses, such as an investigation of the area available for agriculture in the light of demand for land for housing, employment, water management and nature development, modules for the non-urban domain have to be added in the framework. It seems wise to start with the development of a number of modules that contain meta-data from the current configuration of Land Use Scanner and to gradually improve on this by designing a more satisfactory classification of agricultural land use, adding objects such as farms and greenhouses, and

adding attribute information such as landscape management and environmental pressure. The same goes for nature management. In the Spatial Impressions project, a very detailed elaboration of nature management using highly detailed classes and allocation rules was designed that proved to be difficult to put into practice because of a lack of appropriate data (Borsboom-van Beurden et al., 2005). However, evaluation of new nature developments in the National Ecological Network showed that the possibilities for the realisation of specific nature types mainly depends upon local biophysical conditions. For that reason, later configurations as used in the Sustainable Netherlands study simplified this approach, but their oversimplification hampered an assessment of these future land-use changes in terms of biodiversity. Therefore a moderate approach is advised when adding a module on nature management to the framework.

3. *Design of a new real-estate and land-market module*
The current Tigris XL model contains a very basic real-estate and land module, which computes changes in housing supply or land availability for a specific function on the basis of existing land use and a coarse representation of different planning regimes. The incorporation of the cell level in the framework enables much more information from GIS-data at the grid cell level to be input, at the grid-cell level the characteristics of supply of objects and land and characteristics of the environment determining the suitability or attractiveness for specific functions are known. The GIS-data at grid cell level are then aggregated and used to provide information on the availability and suitability of objects and land in a zone or municipality. Thus, at the grid-cell level the characteristics of supply of objects and land and characteristics of the environment determining the suitability or attractiveness for specific functions are known. The characteristics of the demand for objects and land, and demand/supply ratios from the past year, can be derived from the other modules in Tigris XL. In the real-estate and land module, all this information is combined in order to assess the match between the supply of and demand for locations, leading to changes in both the use of objects and land.
4. *Design of an allocation procedure at the cell level*
An efficient allocation procedure needs to be established at the cell level to translate the changes computed at higher levels (municipalities, zones) to specific cells. This procedure can be largely based on the current software used in Land Use Scanner. Intelligence-based, calibrated, transition rules should be capable of delivering a better performance at a higher level than a random-based process. Nevertheless, it will be impossible to make precise forecasts for a specific cell: the outcomes need to be interpreted at a higher scale. Further, the new set-up implies the modelling of changes, not of current land use, in contradiction with the current Land Use Scanner. The incremental structure of the new modelling framework guarantees that most of the land use, as in reality, remains unchanged and that only the relatively small changes are processed. This calculation mechanism at the cell level should also ensure consistency between the different layers, such as actors, objects and land, in the modelling. The large number of cells when a spatial resolution of 100 m $\times$ 100 m is used and the

maintenance of the links between all data layers representing actors, objects and land make computation time and data management at this level important aspects in selecting the right approach.

5. *Integration of the characteristics of supply in other modules*
   The improved information on building and land supply characteristics as made available in the new real-estate and land module described under Point 3 should be included in the housing market, labour market and agriculture modules. The supply information can be used to improve the quality of existing explanatory variables in sector models, to introduce new variables in parameter estimations, or to detail the options for actors by assigning more attributes (such as houses versus houses by type). The advantage of this integration is that the supply of objects and land can play a larger part in the spatial choices in sector models. This would imply, then, that a re-estimation of the modules is needed if new variables or a more detailed set of options are added.
6. *Link with new (or light) version of the National Transport Model*
   The National Transport Model (LMS) of the Ministry of Public Works, Transport and Water Management is an integrated part of the TIGRIS XL framework. It is the ambition of the Ministry and PBL to maintain this set-up and to integrate the new version LMS 2010 in the framework. However, it is expected that the transport module will require the most computation time of all modules. For applications requiring a less detailed focus on transport, the inclusion of a light version of the new LMS is considered as it could be very beneficial for reducing total computation time.
7. *Continuation of the work on water indicators*
   In 2009, a better connection between the current version of Land Use Scanner and DamageScanner (De Bruijn, 2008) was realised by PBL and Deltares. This was done to improve the use of land-use modelling outcomes for the computation of indicators describing the exposure to flooding risks. Alternative spatial developments can be evaluated based on the potential risks they create for flood damage. In the coming period, other water management themes of interest need to be further integrated with land-use changes. Major issues are the lack of water or a local surplus of water and flooding, related to consequences of climate change as intensified periods of drought and rainfall. Another major issue is the quality of water, in relation to the protection of biodiversity and interactions between agriculture and nature. This integration can be done by calculating in a post-processing step the impacts of alternative land-use plans on water indicators, and by using information on the water themes in the design of spatial alternatives.

After a prototype of the new model – incorporating the new framework – has been developed, it is important to test the new system thoroughly by applying it to diverse 'real life' policy questions or scenario explorations. The test phase should measure the performance of the system and generate a list of priorities for further development.

11.7 Conclusion

Since 1997, both Land Use Scanner and Environment Explorer have made valuable contributions to a wide range of policy assessment studies carried out. However, now, nearly 15 years after their development began, progress in computing power, in the availability of detailed data (in particular GIS-data) and in the methods and techniques available for the modelling of human activities have opened up a wide range of new possibilities in this field. These changes have shed new light on the desired performance and sophistication of Land Use Scanner and Environment Explorer. At the same time, a shift in policy questions since the 1990s, as charted by Kuijpers-Linde and Koomen (2009), has prompted the need for the availability of reliable outcomes of land-use modelling at lower scales and for more attention to be paid to the state and management of the environment in its broadest sense.

The PBL initiative to include the Tigris XL model in the new LUMOS model framework addresses these new challenges and opportunities. The new system derives its strength from the combination of modelling expertise from different modelling backgrounds: actor-driven demand modelling and the integrated modelling of the urban and regional system is adopted from the Land Use and Transport Interaction models and this expertise is combined with detailed land-use data and modelling from a cell-based land-use model. The expertise of specific modules, such as demography, housing, labour, transport and agriculture, is further enriched within the context of the overall framework by knowledge and information derived from sector specific models in these fields.

It is clear that the road ahead in developing a new LUMOS model will not always be easy and undoubtedly there are many pitfalls to be avoided, but all constituents required for its successful development by PBL and its partners are present. We therefore hope that this endeavour, which represents a truly collaborative effort, will be broadly welcomed, supported and rewarded with the success it deserves.

References

Atzema, O., Korteweg, P., Lambooy, J., & Van Oort, F. (2009). *Factoren achter het ruimtegebruik van werken. Een literatuurverkenning*. Utrecht: Utrecht University.

Borsboom, J., & Zondag, B. (2009). *Visiedocument modelontwikkeling voor ruimtegebruiks-modellering bij het PBL*. Bilthoven: Internal note, PBL.

Borsboom-van Beurden, J. A. M., Boersma, W. T., Bouwman, A. A., Crommentuijn, L. E. M., Dekkers, J. E. C., & Koomen, E. (2005). *Ruimtelijke Beelden; Visualisatie van een veranderd Nederland in 2030*. RIVM report 550016003. Bilthoven: Milieu- en Natuurplanbureau.

Bröcker, J. (2004). Computable general equilibrium analysis in transportation economics. Chapter 12. In D.A. Hensher, K.J. Button, K.E. Haynes, & P.R. Stopher (Eds.), *Handbook of transport geography and spatial systems* (pp. 269–289). Elsevier, Amsterdam.

CPB, MNP and RPB (2006). *Welvaart en Leefomgeving*. Een scenariostudie voor Nederland in 2040. The Hague: Centraal Planbureau, Milieu- en Natuurplanbureau en Ruimtelijk Planbureau.

De Bruijn, K. (2008). *Bepalen van schade ten gevolge van overstromingen. Voor verschillende scenario's en bij verschillende beleidsopties*. Deltares, Utrecht: Report in commission of IVM.

De Graaff, T., Van Oort, F. G., & Florax, R. J. G. F. (2009). *Sectoral differences in regional population – employment dynamics in the Netherlands*. Paper prepared for the North American Regional Science conference, San Francisco.

De Niet, R., & Van Middelkoop, M. (2009). *Drijvende krachten achter recreatie en toerisme. Quickscan voor LUMOS 1.0. Internal note*. The Hague/Bilthoven: Planbureau voor de Leefomgeving.

Dekkers, J. E. C., & Koomen, E. (2006). *De rol van sectorale inputmodellen in ruimtegebruiksimulatie; Onderzoek naar de modellenketen voor de LUMOS toolbox. SPINlab research memorandum SL-05*. Amsterdam: Vrije Universiteit.

Delta Committee. (2008). Working together with water. A living land build for its future. Findings of the Delta Commissie 2008. Summary and conclusions.

Ettema, D., Floor, H., & De Jong, T. (2009). *Behavioural mechanisms in land use change models. Literature scan*. Utrecht: Faculty of Geosciences/Utrecht University.

Groot, A., Moors, E., Vos, C., Vellinga, P., & Opdam, P. (2009). *Drijvende krachten achter veranderingen in ruimtegebruik voor de sectoren water en natuur, rapport in opdracht van PBL*. Wageningen: University of Wageningen.

Guhathakurta, S. (2003). Advances in urban and environmental modeling: Surveying the terrain and demarcating frontiers. In S. Guhathakurta (Ed.), *Integrated land use and environmental models. A survey of current applications and research* (pp. 3–12). Berlin: Springer.

Hilferink, M., & Grashoff, P. (2009). ICT aspecten Lumos 2.0 ontwikkeling. Object Vision and Demis.

Jacobs, C. (2010). *De drijvende kracht van mobiliteit. Achtergrondstudie bij de vervaardiging van een nieuw LUMOS model*. Amsterdam: VU University.

Koomen, E., Kuhlman, T., Groen, J., & Bouwman, A. A. (2005). Simulating the future of agricultural land use in The Netherlands. *Tijdschrift voor Economische en Sociale Geografie (Journal of Economic and Social Geography), 96*(2), 218–224.

Kuijpers-Linde, M., & Koomen, E. (2009). *Beleidsvragen en indicatoren voor een nieuw ruimtegebruiksmodel, In opdracht van het Planbureau voor de Leefomgeving*. Amsterdam: Geodan Next.

Ligtenberg, A. (2006). *Exploring the use of multi-agent systems for interactive multi-actor spatial planning*. Wageningen: Thesis Wageningen University.

Miller, E. J., Kriger, D. S., & Hunt, J. D. (1998). Integrated urban models for simulation of transit and land-use policies. Final Report TCRP Web Document 9, University of Toronto.

MNP (2007). *Nederland Later; Tweede Duurzaamheidsverkenning deel fysieke leefomgeving Nederland*. MNP-publicatienr.500127001/2007. Bilthoven: Milieu- en Natuurplanbureau.

MNP (2004a). *Quality and the future*. Sustainability outlook. Bilthoven: Milieu- en Natuurplanbureau.

MNP (2004b). Milieu- en Natuureffecten Nota Ruimte, RIVM-rapport 711931009, RIVM, Bilthoven.

Oosterhaven, J., Knaap, T., Ruigrok, C., & Tavasszy, L. (2001). *On the development of RAEM: the Dutch spatial general equilibrium model and its first application to a new railway link*. Proceedings 41st Congress of the European Regional Science Association, Zagreb, Croatia.

PBL (2010). *The Netherlands in the future. Second sustainability outlook: The physical living environment in the Netherlands*. Bilthoven: Netherlands Environmental Assessment Agency (PBL).

Priemus, H., & Hoekstra, J. (2009). Drijvende krachten achter woningvraag en ruimtegebruik wonen. Achtergrondstudie in opdracht van het Planbureau voor de Leefomgeving. Onderzoeksinstituut OTB/Technische Universiteit Delft.

Rijken, B. (2009). *De impact van energie als drijvende kracht van ruimtegebruiksveranderingen*. PBL, Bilthoven: Literature Scan. Internal PBL note.

RIVM (1997). *Nationale Milieuverkenning 1997–2020*. Bilthoven: Rijksinstituut voor Volksgezondheid en Milieu.

RIVM and Stichting DLO. (2001). Who is afraid of red, green and blue? Toets van de Vijfde Nota Ruimtelijke Ordening op ecologische effecten, RIVM-rapportnr. 711931005, Wilco BV, Amersfoort.

RIVM and Stichting DLO. (2002). *National nature outlook 2: 2000–2030*. Summary. Bilthoven: RIVM.

RPB/CBS (2005). Achtergonden en verondertsellingen bij het model PEARL – Naar een nieuwe regionale bevolkings- en allochtonenprognose, NAi Uitgeverij Rotterdam.

Salvini, P., & Miller, E. J. (2005). ILUTE: An operational prototype of a comprehensive microsimulation model of urban systems. *Networks and Spatial Economics, 5*, 217–234.

Simmonds, D. (1999). The design of the Delta land-use modelling package. *Environment and Planning B: Planning and Design, 26*, 665–684.

Simmonds, D. C., & Echenique, M. (1999). DETR report, Review of land use/transport interaction models, London.

Timmermans, H. (2007). Welvaart en Leefomgeving: reflectie op scenario ontwikkeling. Internal report. Technical University Eindhoven.

Van Bruchem, C., & Silvis, H. (2009). *Drijvende krachten toekomstig landgebruik landbouw*. The Hague: LEI.

Van Wissen, L. (2000). A micro-simulation model of firms: Applications of concepts of the demography of the firm. *Papers of Regional Science, 79*(2), 111–134.

Van de Velde, R. J., Schotten, C. G. J., Van der Waals, J. F. M., Boersma, W. T., Ouwersloot, H., & Ransijn, M. (1997). *Ruimteclaims en ruimtelijke ontwikkelingen in de zoekgebieden voor de toekomstige nationale luchtinfrastructuur (TNLI)*. Quickscan met de Ruimtescanner. RIVM-rapport 711901024. Bilthoven: RIVM.

Verburg, P., & Overmars, K. (2009). Combining top-down and bottom-up dynamics in land use modeling: Exploring the future of abandoned farmlands in Europe with the Dyna-CLUE model. *Landscape Ecology, 24*, 1167–1181.

VROM (2008). *Randstad towards 2040, summary of the structural vision*. The Hague: Ministry of Housing, Spatial Planning and the Environment.

Wagner, P., & Wegener, M. (2007). Urban land use, transport and environment models – experiences with an Integrated Microscopic Approach.

Wegener, M. (2004). Overview of land use transport models. Chapter 9. In D.A. Hensher, K.J. Button, K.E. Haynes, & P.R. Stopher (Eds.), *Handbook of transport geography and spatial systems* (pp. 127–146). Elsevier, Amsterdam.

Weidner, T., Donelly, R., Freedman, J., Abraham, J. E., & Hunt, J. D. (2006). TLUMIP – transport and land use model in Portland – current state, Stadt Region Land 81, Aachen: institute fur Stadtbauwesen und stadtverkehr, RWTH Aachen, pp. 91–102.

Zondag, B. (2007). Joint modeling of land-use, transport and economy, PhD thesis TRAIL thesis series nr. T2007/4, Delft University of Technology, Delft.

Zondag, B., & Borsboom, J. (2009). *Driving forces of land-use change*. Paper prepared for the 49th ERSA conference August 2009, Lodz, Poland.

Zondag, B., & Pieters, M. (2001). Literature Review of Land-Use model, prepared by RAND Europe for the Transportation Research Centre of the Netherlands Ministry of Transport, Public Works and Water Management, Leiden.

Index

E. Koomen, J. Borsboom-van Beurden (eds.), *Land-Use Modelling in Planning Practice*, The GeoJournal Library 101, DOI 10.1007/978-94-007-1822-7,

W

Z

Printed by Books on Demand, Germany